CENTRAL NERVOUS SYSTEM CONTROL
OF THE HEART

TOPICS IN THE NEUROSCIENCES

Other books in the series:

Rami Rahamimoff and Sir Bernard Katz, eds.: *Calcium, Neuronal Function and Transmitter Release*. ISBN 0–89838–791–4.

Robert C.A. Frederickson, ed.: *Neuroregulation of Autonomic, Endocrine and Immune Systems*. ISBN 0–89838–800–7.

Giuditta, et al., eds.: *Role of RNA and DNA in Brain Function*. ISBN 0–89838–814–7.

CENTRAL NERVOUS SYSTEM CONTROL OF THE HEART

Proceedings of the
IIIrd International Brain Heart Conference
Trier, Federal Republic of Germany

edited by

T. Stober
University of Saarland
Homburg/Saar, FRG

K. Schimrigk
University of Saarland
Homburg/Saar, FRG

D. Ganten
University of Heidelberg
Heidelberg, FRG

D.G. Sherman
University of Texas Health Science Center
San Antonio, Texas, USA

Martinus Nijhoff Publishing
a member of the Kluwer Academic Publishers Group
Boston / Dordrecht / Lancaster

Distributors for North America:
Kluwer Academic Publishers
101 Philip Drive
Assinippi Park
Norwell, Massachusetts 02061, USA

Distributors for the UK and Ireland:
Kluwer Academic Publishers
MTP Press Limited
Falcon House, Queen Square
Lancaster LA1 1RN, UNITED KINGDOM

Distributors for all other countries:
Kluwer Academic Publishers Group
Distribution Centre
Post Office Box 322
3300 AH Dordrecht, THE NETHERLANDS

Library of Congress Cataloging-in-Publication Data

International Brain Heart Conference (3rd : 1985 : Trier,
  Germany)
  Central nervous system control of the heart.

  (Topics in the neurosciences)
  Includes bibliographies and index.
  1. Heart—Innervation—Congresses. 2. Nervous
system, Vasomotor—Congresses. 3. Cerebrovascular
disease—Congresses. I. Stober, T. II. Title.
III. Series. [DNLM: 1. Cardiovascular Diseases—
physiopathology—congresses. 2. Cardiovascular
System—physiology—congresses. 3. Central Nervous
System—physiology—congresses. 4. Central Nervous
System Diseases—physiopathology—congresses.
5. Cerebral Embolism and Thrombosis—physiopathology—
congresses. W3 IN1246 3rd 1985c / WC 100 I565 1985c]
QP113.4.I57 1985   612'.178   86-16331
ISBN 0-89838-820-1

CONTENTS

PART FOUR: CARDIOEMBOLIC STROKE

CONTRIBUTING AUTHORS

AGNATI, L.F.
Department of Human Physiology
University of Modena
Modena/Italy

ANDREOLI, A.
Department of Neurosurgery
Bellaria Hospital
Via Altura 3
I-40139 Bologna/Italy

ANSTÄTT, T.
Neurologische Klinik der
Universitätskliniken Homburg
D-6650 Homburg, Saar/FRG

AVERILL, D.B.
Research Division
Cleveland Clinic Foundation
9500 Euclid Avenue
Cleveland, Ohio 44106/USA

BARNES, K.L.
Research Division
Cleveland Clinic Foundation
9500 Euclid Avenue
Cleveland, Ohio 44106/USA

BARNETT, H.J.M.
University Hospital
P.O. Box 5339
London, Ontario N6A 5A5/Canada

BIRO, G.
Neurologische Klinik der
Universitätskliniken Homburg
D-6650 Homburg, Saar/FRG

BJORK, R.J.
Department of Neurology
University of Minnesota
St. Paul-Ramsey Medical Center
640 Jackson Street
St. Paul, Minnesota 55101/USA

BONNET, J.
Clinique Medicale Cardiologique
Hopital Cardiologique de Bordeaux
F-33604 Pessac/France

BOUGHNER, D.R.
University Hospital
P.O. Box 5339
London, Ontario N6A 5A5/Canada

BRICAUD, H.
Clinique Medicale Cardiologique
Hopital Cardiologique de Bordeuax
F-33604 Pessac/France

CHOKI, J.
Cerebrovascular Division
National Cardiovascular Center
5-7-1 Fujishiro-dai
Suita, Osaka 565/Japan

CIPOLLE, R.J.
Department of Neurology
University of Minnesota
St. Paul-Ramsey Medical Center
640 Jackson Street
St. Paul, Minnesota 55101/USA

CIRIELLO, J.
Depts of Physiology, Clinical
Neurological Sciences and
Pathology
Health Science Center
University of Western Ontario
London, Ontario, N6A 5C1/Canada

CLEMENTY, J.
Clinique Medicale Cardiologique
Hopital Cardiologique de Bordeaux
F-33604 Pessac/France

CONLAY, L.A.
Dept of Anesthesia
Massachusetts General Hospital and
Harvard Medical School
Fruit Street
Boston, Massachusetts 02114/USA

CONOMY, J.P.
Research Division
Cleveland Clinic Foundation
9500 Euclid Avenue
Cleveland, Ohio 44106/USA

COSTE, P.
Clinique Medicale Cardiologique
Hopital Cardiologique de Bordeaux
F-33604 Pessac/France

DESBORDES, P.
Unite de Pathologie
Vasculaire Cerebrale
Hopital Pellegrin
Place Amelie-Raba-Leon
F-33076 Bordeaux cedex/France

DI PASQUALE, G.
Service of Cardiology
Bellaria Hospital
Via Altura 3
I-40139 Bologna/Italy

ERBEL, R.
Second Department of
Internal Medicine
University Hospital of Mainz
Langenbeckstr. 1
D-6650 Mainz/FRG

FADEN, A.I.
Neurology Service (127)
San Francisco Veterans
Administration Medical Center
4150 Clement Street
San Francisco, CA 94121/USA

FERRARIO, C.M.
Research Division
Cleveland Clinic Foundation
9500 Euclid Avenue
Cleveland, Ohio 44106/USA

FREIER, G.
Department of Neurology
Faculty of Medicine
University of the Saarland
D-6650 Homburg, Saar/FRG

FUXE, K.
Department of Histology
Karolinska Institute
Box 60400
S-10401 Stockholm/Sweden

GANTEN, D.
Deutsches Institut zur Bekämpfung
des hohen Blutdruckes und Pharma-
kologisches Institut der
Universität Heidelberg
Im Neuenheimer Feld 366
D-6900 Heidelberg/FRG

GIBSON, C.J.
Departments of Physiology,
Clinical Neurological Sciences
and Pathology
Health Science Center
University of Western Ontario
London, Ontario N6A 5C1/Canada

GOLDSTEIN, M.
Department of Psychiatry
New York University Medical Center
New York/USA

GRANATA, A.
Laboratory of Neurobiology
Department of Neurology
Cornell University Medical College
1300 York Avenue
New York, N.Y. 10021/USA

HACHINSKI, V.C.
University Hospital
of Western Ontario
P.O. Box 5339
Station "A"
London, Ontatio N6A 5A5/Canada

HÄRFSTRAND, A.
Department of Histology
Karolinska Institute
Box 60400
S-10401 Stockholm/Sweden

HAMANN, G.
Department of Neurology
Faculty of Medicine
University of the Saarland
D-6650 Homburg, Saar/FRG

HART, R.G.
Department of Neurology
University of Texas
Health Science Center
7703 Floyd Curl Drive
San Antonio, Texas 78284/USA

HERMAN, B.
Institute of Neurology
Catholic University
Reiner Postlaan 4
NL-6500 HB Nijmegen/
The Netherlands

HORNIG, C.R.
Neurologische Universitätsklinik
Am Steg
D-6300 Giessen/FRG

KALIA, M.
Department of Pharmacology
Thomas Jefferson University
Medical Center
Philadelphia/USA

KOWITZ, J.
Department of Neurology
University of Minnesota
St. Paul-Ramsey Medical Center
640 Jackson Street
St. Paul, Minnesota 55101/USA

KRÄMER, G.
Department of Neurology
University Hospital of Mainz
Langenbeckstr. 1
D-6500 Mainz/FRG

KURAMOTO, K.
Tokyo Metropolitan
Geriatric Hospital
35-2 Sakaecho
Itabashiku
Tokyo 173/Japan

LANG, R.E.
Deutsches Institut zur Bekämpfung
des hohen Blutdruckes und Pharma-
kologisches Institut der
Universität Heidelberg
Im Neuenheimer Feld 366
D-6900 Heidelberg/FRG

LA VECCHIA, F.
Service of Cardiology
Bellaria Hospital
Via Altura 3
I-40139 Bologna/Italy

LEYTEN, A.C.M.
General Hospital
Tilburg/
The Netherlands

LODDER, J.
Department of Neurology
Medical Faculty
University of Limburg
P.O. Box 616
NL-6200 MD Maastricht/
The Netherlands

LOEWY, A.D.
Department of Anatomy
and Neurobiology
Washington University
School of Medicine
Box 8108
660 South Euclid Avenue
St. Louis, Missouri 63110/USA

LUSA, A.M.
Service of Cardiology
Bellaria Hospital
Via Altura 3
I-40139 Bologna/Italy

MCINTOSH, T.K.
Department of Neurology
University of California
4150 Clement Street
San Francisco, California 94121/USA

MAHER, T.J.
Department of Anesthesia
Massachusetts General Hospital and
Harvard Medical School
Fruit Street
Boston, Massachusetts 02114/USA

MANINI, G.L.
Service of Cardiology
Bellaria Hospital
Via Altura 3
I-40139 Bologna/Italy

MATSUSHITA, S.
Tokyo Metropolitan
Geriatric Hospital
35-2 Sakaecho
Itabashiku
Tokyo 173/Japan

MERKEL, K.H.
Krankenhaus am Urban
Pathologisches Institut
Dieffbachstr. 1
D-1000 Berlin 61/FRG

MINEMATSU, K.
Cerebrovascular Division
National Cardiovascular Center
5-7-1 Fujishiro-dai
Suita, Osaka 565/Japan

MIYASHITA, T.
Cerebrovascular Division
National Cardiovascular Center
5-7-1 Fujishiro-dai
Suita, Osaka 565/Japan

MOHR-KAHALY, S.
Second Dept of Internal Medicine
University Hospital of Mainz
Langenbeckstr. 1
D-6500 Mainz/FRG

NEUMEYER, A.
Department of Histology
Karolinska Institute
Box 60400
S-10401 Stockholm/Sweden

OMAE, T.
Cerebrovascular Division
National Cardiovascular Center
5-7-1 Fujishiro-dai
Suita, Osaka 565/Japan

ORGOGOZO, J.M.
Unite de Pathologie
Vasculaire Cerebrale
Hopital Pellergin
Place Amielie-Raba-Leon
F-33076 Bordeaux cedex/France

PINELLI, G.
Service of Cardiology
Bellaria Hospital
Via Altura 3
I-40139 Bologna/Italy

RAMIREZ-LASSEPAS, M.
Department of Neurology
University of Minnesota
St. Paul-Ramsey Medical Center
640 Jackson Street
St. Paul, Minnesota 55101/USA

REIS, D.J.
Laboratory of Neurobiology
Department of Neurology
Cornell University Medical College
1300 York Avenue
New York, N.Y. 10021/USA

REM, J.A.
Depts of Clinical Neurological
Sciences and Medicine
University Hospital
P.O. Box 5339
London, Ontario N6A 5A5/Canada

ROSSI, E.
Instituti di Cardiologia
dell'Universita Cattolica
del Sacro Cuore
"Agostino Gemelli"
Largo Gemelli 8
I-00168 Roma/Italy

ROSSI, G.F.
Instituti di Neurochirurgia
dell'Universita Cattolica
del Sacro Cuore
"Agostino Gemelli"
Largo Gemelli 8
I-00168 Roma/Italy

RUGGIERO, D.A.
Laboratory of Neurobiology
Department of Neurology
Cornell University Medical College
1300 York Avenue
New York, N.Y. 10021/USA

SCHIMRIGK, K.
Neurologische Klinik der
Universitätskliniken Homburg
D-6650 Homburg, Saar/FRG

SCHULTE, B.P.M.
Institute of Neurology
Reiner Postlaan 4
Postbus 9101
NL-6500 HB Nijmegen/
The Netherlands

SEN, S.
III. Med. Klinik der
Universitätskliniken Homburg
D-6650 Homburg, Saar/FRG

SHAPIRO, D.
Department of Psychiatry
University of California
760 Westwood Plaza
Los Angeles, California 90024/USA

SHERMAN, D.G.
Department of Neurology
University of Texas
Health Science Center
7703 Floyd Curl Drive
San Antonio, Texas 78284/USA

SILVER, M.D.
Depts of Physiology, Clinical
Neurological Sciences and
Pathology
Health Science Center
University of Western Ontario
London, Ontario N6A 5C1/Canada

SMITH, K.E.
Depts of Physiology, Clinical
Neurological Sciences and
Pathology
University of Western Ontario
London, Ontario N6A 5C1/Canada

SNYDER, B.D.
Department of Neurology
St. Paul-Ramsey Medical Center
640 Jackson Street
St. Paul, Minnesota 55101/USA

STEIN, S.D.
Department of Neurology
St. Paul-Ramsey Medical Center
640 Jackson Street
St. Paul, Minnesota 55101/USA

STOBER, T.
Neurologische Klinik der
Universitätskliniken Homburg
D-6650 Homburg, Saar/FRG

SUTER, T.W.
Department of Psychiatry
University of California
760 Westwood Plaza
Los Angeles, California 90024/USA

SVENSSON, T.H.
Department of Pharmacology
Karolinska Institute
Box 60400
S-10401 Stockholm/Sweden

TASHIRO, M.
Cerebrovascular Division
National Cardiovascular Center
5-7-1 Fujishiro-dai
Suita, Osaka 565/Japan

TENCATI, R.
Service of Cardiology
Bellaria Hospital
Via Altura 3
I-40139 Bologna/Italy

TERENIUS, L.
Department of Pharmacology
University of Uppsala
Uppsala/Sweden

TOGNETTI, F.
Department of Neurosurgery
Bellaria Hospital
Via Altura 3
I-40139 Bologna/Italy

TOPHOF, M.
Department of Neurology
University Hospital of Mainz
Langenbeckstr. 1
D-6500 Mainz/FRG

UNGER,Th.
Deutsches Institut zur Bekämpfung
des hohen Blutdruckes und Pharma-
kologisches Institut der
Universität Heidelberg
Im Neuenheimer Feld 366
D-6900 Heidelberg/FRG

WEBER, J.C.
Department of Neurology
University of Minnesota
St. Paul-Ramsey Medical Center
640 Jackson Street
St. Paul, Minnesota 55101/USA

WEIPERT, D.
Department of Psychiatry
University of California
760 Westwood Plaza
Los Angeles, California 90024/USA

WURTMAN, R.J.
Department of Anesthesia
Massachusetts General Hospital and
Harvard Medical School
Fruit Street
Boston, Massachusetts 02114/USA

YAMAGUCHI, T.
Cerebrovascular Division
National Cardiovascular Center
5-7-1 Fujishiro-dai
Suita, Osaka 565/Japan

YAMANOUCHI, H.
Tokyo Metropolitan
Geriatric Hospital
35-2 Sakaecho
Itabashiku
Tokyo 173/Japan

ZENKER, G.
Second Dept of Internal Medicine
LKH, Graz; and
Institute of Preventive Medicine
Joanneum/Austria

ZOLI, M.
Department of Human Physiology
University of Modena
Modena/Italy

PREFACE

The first two "Brain Heart Conferences" in Jerusalem in 1978 and 1983 were based upon the common interests of clinically orientated neurologists and cardiologists in the problems of central autonomic control and autonomic disturbances of the cardiovascular system. The relatively slow scientific progress, at least clinically, in this area may be due to the fact that neither cardiologists nor neurologists felt competent in both topics. Furthermore, it has become increasingly difficult to have an overall view of the basic research and its clinical applications in this field. New research methods, based on a combination of morphological, biochemical, and physiological techniques, have enabled the functional differentiation of various areas of the brain and subsequently also of the autonomic nervous system. The simple dualistic concept of an antagonistic sympathetic-parasympathetic regulation of the circulatory system is no longer valid. It is clear that numerous neurotransmitters, in particular the neuropeptides, are involved in a highly differentiated subdivision of the autonomic system.

One of the aims of the IIIrd International Brain Heart Conference was therefore to supply a synopsis of the latest developments in basic research undertaken in this field by exceptionally competent scientists, to clinically orientated neurologists and cardiologists, and thus to provide new impulses for clinical research.

Clinical concepts of cardioembolic stroke are rudimentary and scientifically flawed. The Cardioembolic Stroke Workshop brought together, for the first time, an international group of clinicians actively working in this difficult area. The Workshop provided a forum to define important issues, to share hypotheses, and to plan better clinical studies. Progress in the understanding of cardioembolic stroke will depend on the cooperation of cardiologists and neurologists to design methodologically sound studies of the important interrelationship of the heart and the brain. The clinical science of cardioembolic stroke took a small step forward at the workshop.

The resulting book shows that these aims have been achieved through the active participation of all concerned. Our special thanks go to the authors for their excellent contributions; to Mr. J. Stier from the Martinus Nijhoff Publishing Company for his support in this project and the rapid publication of the congress proceedings; to all members of the scientific committee; and last but not least to Ms. H. Schnöring and Ms. S. Ruby for the typing of the manuscripts, to Ms. C. Ullmann and Dr. R. Kiehl for the linguistic revision of some articles and to Mr. M. Dietsche for the proof-reading.

Homburg/Saar, Heidelberg, San Antonio

T. Stober   D. Ganten
K. Schimrigk   D.G. Sherman

PART ONE

ANATOMICAL AND PHYSIOLOGICAL
ASPECTS OF CENTRAL CARDIOVASCULAR CONTROL

1

ANATOMIC ASPECTS OF CENTRAL NERVOUS CARDIOVASCULAR REGULATION

A.D. LOEWY

Department of Anatomy and Neurobiology, Washington University School of Medicine, St. Louis, Missouri, USA

ABSTRACT

Cardiovascular afferent information is relayed from baro- and chemoreceptors located in the aortic arch and carotid sinus to the nucleus tractus solitarius (NTS). This information is then relayed to a number of brainstem and forebrain nuclei. Two important concepts emerge from connectional studies of the NTS efferent pathways. First, the NTS projects directly to some of the brainstem and hypothalamic nuclei which give rise to direct pathways that control either vagal or sympathetic preganglionic neurons. For example, the control over the sympathetic outflow can be mediated by the A5 noradrenergic cell group, the ventral medulla which includes the C1 adrenergic cell group as well as the descending substance P pathway, the Kolliker-Fuse region of the parabrachial nucleus, and the paraventricualr nucleus. Second, the NTS can modulate the release of vasopressin by means of indirect circuits. Some of these includes pathways from the A1 and A2 cell groups as well as the lateral parabrachial nucleus.

In the last 10 years, with the development of highly sensitive neuroanatomical methods (e.g., anterograde or retrograde axonal transport methods and the immunohistochemical technique which allows biochemical definition of neural pathways), there has been a significant advancement in cur knowledge of the central neural circuits which are involved in regulation of autonomic functions. One of the ideas to emerge from this work is that there is an interrelationship between those central neural pathways which regulate preganglionic motor neurons and those that are involved in regulation of vasopressin release from the hypothalmus. This body of neuroanatomical information has advanced far beyond our current understanding of the physiological mechanisms of either of the two systems. Nevertheless, it provides an important conceptual framework for future work. The objective of this paper is to summarize some of this information.

CARDIOVASCULAR AFFERENT CONNECTIONS

Visceral afferent information from the heart and its great vessels is conveyed into the CNS by two neural systems: dorsal root and vagal afferents. The afferent dorsal root fibers that travel with the sympathetic fibers carry sensory information from the heart and blood vessels to the dorsal horn of the spinal cord. This system conveys both tonic cardiovascular and nociceptive information. Various functional classes of sympathetic afferents have been identified that innervate the atria, ventricles, coronary blood vessels, and other systemic blood vessels (see 1 for review). In fact, the entire cardiovascular system including both arteries and veins of all diameters is innervated by a substance P-like set of nerves originating from the dorsal root ganglia (2). While details regarding the central circuits involved in processing this spinal afferent information is still sketchy, it appears that the visceral afferent fibers project to specific neurons in laminae I and V of the spinal cord, some of these neurons also receive somatic inputs (3). This convergence of visceral and somatic afferent information onto spinal cord relay neurons is part of the neural mechanism responsible for referred pain (Figure 1). This information from laminae I and IV neurons is relayed to the medial thalamus (4) and to the parabrachial nucleus (5). In turn, it is likely that this information is transmitted to the insular cortex (Figure 2) because this is the major cortical area receiving projections from these two regions (6). Just how this central afferent system is involved in regulation of the autonomic nervous system is unknown, but this area of the cortex provides a direct pathway to the parasympathetic preganglionic nuclei and thus, may modulate the cardioinhibitory system (7).

The nucleus tractus solitarus (NTS) is the main afferent relay station for cardiovascular information (as well as other visceral information).

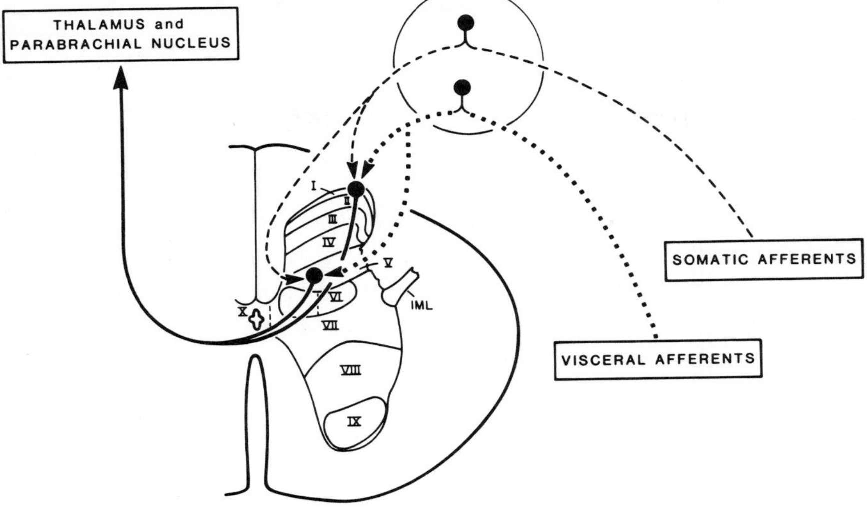

Fig. 1. Visceral and somatic sensory information is relayed to laminae
I and V of the dorsal horn. Some of this information converges
on the same neuron and is probably related to the parabrachial
nucleus and/or medial thalamic nuclei.

Afferent fibers from baroreceptors and chemoreceptors of the carotid sinus
and aortic arch travel respectively in the IXth and Xth cranial nerves to
the NTS. The exact site of termination within the NTS that receives this
information has been determined using the transganglionic horseradish
peroxidase technique (8,9). This technique works as follows: The distal
branch of a nerve (e.g. carotid sinus nerve) is isolated from its end
organ, incubated in a solution of horseradish peroxidase for 2 days. The
enzyme is transported intraaxonally from the distal site, through the
cytoplasm of the cell body, and into the terminal portions of proximal
axonal branches which terminate in specific loci in the NTS. The animal is
perfused with fixative, and then the brain sections are processed
histochemically to stain for horseradish peroxidase that is transported in
the nerve. This technique, however, does not allow identification of any
specific physiological class of axon and thus has the disadvantage of

failing to distinguish between baroreceptor and chemoreceptor fibers. In order to get around this problem, Wallach and Loewy (9) studied the central projections of the aortic nerve in the rabbit which is a pure barosensory nerve. This nerve projects mainly to the dorsolateral NTS area with lighter projections to the medial and commissural nuclei. Using electrophysiological techniques, Donoghue et al. (10) have also shown that three areas are main sites in the NTS which receive a baroreceptor input from both the carotid sinus and aortic arch nerves. Chemoreceptor fibers project mainly to the medial and commissural regions of the NTS and appear to project somewhat more caudally in the commissural nucleus than the baroreceptor fibers. Thus, their data suggest that there is convergence of cardiovascular information in the NTS.

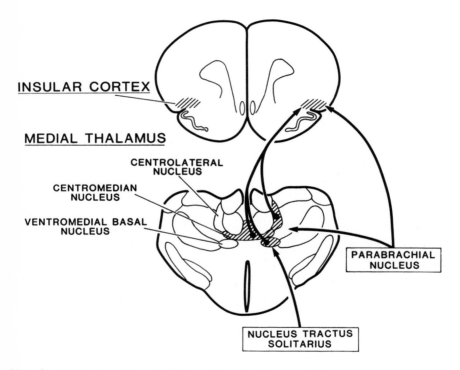

Fig. 2. Visceral afferent information projects to the insular cortex via two major routes: one through the intralaminar nuclei of the medial thalamus and the other by the relay from the parabrachial nucleus.

The question of what is the neurotransmitter that mediates the baroreceptor reflex at the first synapse in the NTS has been studied. Haeusler and Osterwalder (11) suggested that substance P was a potential neurotransmitter by showing that application of this peptide or capsaicin (the active ingredient in red peppers - that causes release of substance P - as well possible neuropeptides) on the dorsal medulla just above the NTS elicited a decrease in blood pressure and heart rate mimicking the baroreceptor reflex. Anatomical studies by Gillis et al. (12) indicate that substance P-immunoreactivity is present in nerve fibers that innervate the carotid sinus and aortic arch, in the cell bodies of the nodose and petrosal ganglia, and in the NTS. The substance P immunoreactivity in the NTS decreases after intracranial section of the IXth and Xth cranial nerves. Their findings support the hypothesis presented by Haeusler and Osterwalder. However, Furness and associates (13) demonstrated that the baroreceptor reflex was unaffected after capsaicin induced depletion of substance P-containing nerves which innervate the heart and blood vessels. Similarly, Talman and Reis (14) reported that microinjections of low doses of substance P into the NTS did not elicit a fall in blood pressure or heart rate. They attributed the fall in blood pressure seen after high doses (which required an increased volume of injectate because of limits of solubility of the peptide) due to local distortion of the NTS.

Talman and Reis (14) reported that injections of the excitatory amino acid, L-glutamate, into the NTS cause bradycardia and hypotension; this effect is similar to the baroreceptor reflex. While this finding is, by itself, not particularly surprising in view of the fact L-glutamate excites almost all cells in the CNS, microchemical studies of the NTS have demonstrated the high presence of L-glutamate in the specific regions of the NTS that receive a primary afferent input from the baroreceptor nerves (15). Transection of the IXth and Xth cranial nerves caused a decrease in L-glutamate in these NTS regions. A high affinity uptake system for L-glutamate has been demonstrated in the NTS by Talman et al. (16) and Perrone (17). While these results may represent changes in a glutamate synaptic mechanism, there is the possibility that they reflect changes in metabolism. Therefore, a critical experiment that is still needed is the demonstration with combined immunocytochemical and electron microscopic techniques that L-glutamate is present in synaptic terminals of primary afferent fibers in the dorsolateral region of the NTS.

EFFERENT CONNECTIONS OF THE NUCLEUS TRACTUS SOLITARIUS

The main efferent projections of the NTS are summarized in figure 3. This drawing is based on the autoradiographic studies published by Ricardo and Koh (18) and Loewy and Burton (19). Most of the NTS projections are bilateral, but for the sake of graphic clarity, only the ipsilateral connections are illustrated. An important point to bear in mind when examining Figure 3 is that the NTS is involved in a wide range of visceral functions, ranging from taste, cardiovascular activity, respiratory control, to gastric functions and this type of anatomical analysis does not provide any information regarding specific visceral functions being subserved at these different loci in the CNS. Furthermore, the NTS is composed of at least six different subnuclei which all have their own set of unique connections and functions (see 20 and 21 for recent reviews). Thus, our current understanding of the efferent connections of the NTS is not yet detailed enough to provide a comprehensive discussion of specific autonomic functions. Nevertheless, two important concepts regarding central cardiovascular function have emerged from the neuroanatomical studies. First, the NTS has connections with a number of regions in the CNS which either directly or indirectly control preganglionic activity. For example, there is a NTS-nucleus ambiguus pathway which controls the inhibitory vagal outflow to the heart. In addition, it provides dense projections to various brain stem nuclei which provide secondary projections to the sympathetic preganglionic cell column which includes the A5 noradrenergic cell group (22), Kölliker-Fuse nucleus (23), the ventral medulla (24), and the C1 adrenergic cell group. Second, the NTS by way of its connections with A1 cell group (25) or via pathways to the parabrachial nucleus (18,19) provide ascending circuits that are likely to be involved in regulating the release of vasopressin. These pathways will be discussed below.

One of these circuits that is of critical importance in regulating blood pressure is the baroreceptor pathway. The proposed circuitry of this reflex is illustrated in figure 4. Stretch receptors from the aortic arch and carotid sinus send fibers mainly to the dorsolateral NTS subnucleus. While it has not yet been demonstrated, it is quite likely that the dorsolateral nucleus connects via interneurons with the medial and

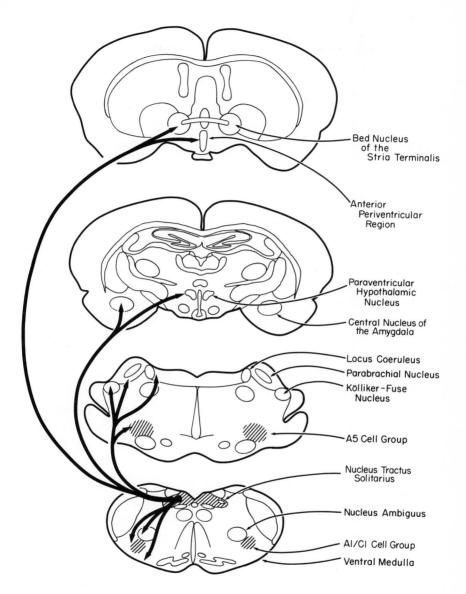

Bed Nucleus
of the
Stria Terminalis

Anterior
Periventricular
Region

Paraventricular
Hypothalamic
Nucleus

Central Nucleus of
the Amygdala

Locus Coeruleus
Parabrachial Nucleus
Kölliker-Fuse
Nucleus

A5 Cell Group

Nucleus Tractus
Solitarius

Nucleus Ambiguus

Al/Cl Cell Group
Ventral Medulla

Figure 3: Summary of the major efferent connections of the nucleus
tractus solitarius.

commissural subnuclei. The reason for this speculation is the dorsolateral subnucleus has no known efferent projections. In contrast, the medial and commissural nuclei provide numerous efferent connections to brainstem and forebrain areas (18).

One of the critical pathways in this reflex is the NTS projection to the ventral medulla (19). This latter area is the probable site of the vasomotor center. There is considerable evidence that pharmacological manipulation of this region produces profound changes in blood pressure. For example, Yamada, McAllen & Loewy (26) found that application of GABA antagonists to the ventral medulla blocked the baroreceptor reflex in a dose dependent manner. From these data it was deduced that a GABA interneuron may exist at this site. As shown in figure 4, the function of this interneuron is likely to be to provide inhibition of the descending excitatory medullary neurons which control the level of activity of the sympathetic outflow. We have hypothesized that this descending pathway, which is the final common pathway controlling sympathetic vasomotor preganglionic neurons, involves substance P neurons (27). Evidence for this hypothesis is derived from the observation that 1) there is a descending substance P pathway from the ventral medulla to the intermediolateral cell column (24), 2) stimulation of this area leads to both a pressor response which can be blocked by intrathecal injections of substance P antagonists (27), 3) a release of immunoreactive substance P from the spinal cord occurs after cell bodies in the ventral medulla are excited by kainic acid (Takano, Martin, Leeman & Loewy, 1984), and 4) iontophoretic application of substance P on sympathetic preganglionic neurons causes a long-acting excitation (28,29).

NEURAL CIRCUITS INVOLVED IN REGULATION OF VASOPRESSIN RELEASE

It is now established that arginine vasopressin (AVP) - a peptide hormone secreted into the blood stream from the magnocellular hypothalamic nuclei acts as both an antidiuretic hormone and a vasoconstrictor hormone. While it has been known ever since this peptide was first discovered that it had potent vasoconstrictor actions, it has been only within the last few years that strong evidence has been obtained that AVP acts as a potent vasoconstrictor hormone in the normal physiological range (0.3 to 30.0 pg/ml of plasma). Cowley and his associates (30) have summarized their data

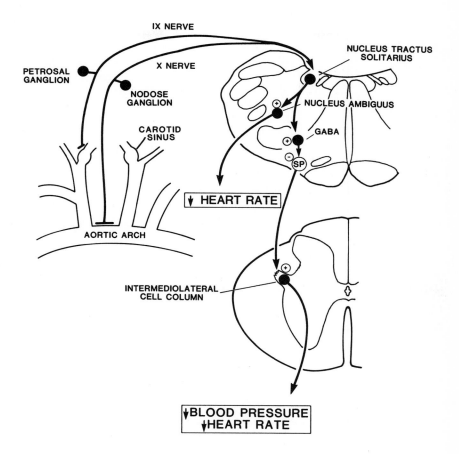

Figure 4: This diagram outlines the neural pathways thought to be involved in the baroreceptor reflex. Incoming baroreceptor
afferent fibers synapse in the nucleus tractus solitarius. The
nucleus tractus solitarius sends one projection to the cardioinhibitory neurons in the nucleus ambiguus and a second projection to the ventral medulla. The latter projection may inhibit the descending medullospinal excitatory neurons -- which
are thought to contain substance P -- by synapsing first on
GABA interneurons.

which show that physiological doses of AVP do not raise arterial blood pressure because reflex changes in cardiac output compensate for the vascular actions of this peptide. Changes in arterial blood pressure can be demonstrated in sinoaortic baroreceptor denervated animals or animals that have had their entire CNS destroyed (30). In addition, AVP has been shown to enhance the gain of the baroreceptor reflex.

Volume receptors localized in the left atrium of the heart provide the main afferent input to the NTS which triggers, through multisynaptic pathways, the release of vasopressin from the supraoptic (SON) and paraventricular nuclei of the hypothalamus (PVN). A decrease in central blood volume (e.g., during hemorrhage) causes a release of AVP. Conversely, when the blood volume is expanded, there is an inhibition in the secretion of AVP. To a lesser degree, chemoreceptors appear to provide facilitory inputs while baroreceptors provide an inhibitory input (31-33) to the vasopressin neurons of the SON. Other factors such as extracellular osmolarity, plasma levels of angiotensin II, and orthostatic reflexes are also involved. The general physiology of these reflexes has been discussed in a number of recent reviews (e.g., 30,34,35) and will not be covered here. The objective here is to briefly review the neural pathways that project to the SON and PVN and to discuss how these might possibly control the release of vasopressin.

Considerable interest has arisen at trying to determine the central neural pathways which are involved in regulating the release of AVP from neurons of the SON and PVN. To date, most of the information that has been gathered deals with the SON. There are three known inputs to the AVP neurons of the SON: A1 noradrenergic cell group, median preoptic nucleus, and subfornical organ. Each of these will be considered below.

For 20 years it has been known that the PVN and SON receive a dense noradrenergic input (36-38). The noradrenergic input to the vasopressin neurons in the SON and PVN appears to arise exclusively from the A1 noradrenergic cell group (Figure 5; 25). The A2 cell group of the NTS projects to the parvocellular division of the PVN -- which is involved in the descending control of the vagal and sympathetic preganglionic neurons and not to the magnocellular divisions of the PVN or SON which contain vasopressin neurons. While it is possible that interneurons from the parvocellular portion of the PVN project to the magnocellular division of the PVN (van den Pol, 1982) and could also be involved in the release of

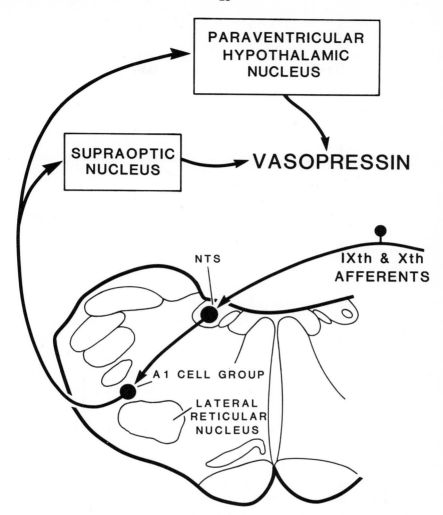

Figure 5: Afferent information from the atria of the heart, baroreceptors, and chemoreceptors is relayed first to the nucleus tractus solitarius. Then, second order neurons project to the A1 cell group. The A1 cell group provides the sole noradrenergic input to magnocellular regions of the supraoptic nucleus and paraventricular hypothalamic nucleus. The noradrenergic pathway from the A2 cell group connects directly with regions in the paraventricular hypothalamic nucleus involved in regulation of the autonomic nervous system.

vasopressin, this has not yet established. None of the other noradrenergic cell groups of the brainstem (A5, A6, A7) provide inputs to the SON or PVN. The locus coeruleus (A6) provides a very light projection to the periventricular area associated with the PVN, but not directly to the neurosecretory cells of either the PVN or SON (39).

The status of noradrenaline as neurotransmitter in the SON and PVN is confusing. Three lines of evidence suggest that noradrenaline acts as an inhibitory transmitter. First, ionophoretic application of noradrenaline on neurosecretory cells in both the PVN and SON causes inhibition of neural activity (40-43). Second, in organ cultures of the hypothalamoneurohypophyseal system, noradrenaline inhibits the release of vasopressin (44). Third, Blessing et al. (45) have demonstrated that lesions of the A1 cell group cause an elevation in plasma vasopressin and blood pressure that is independent of the sympathetic nervous system. On the other hand, experiments have been performed which suggest that noradrenaline acts as an excitatory transmitter in this system. First, central injections of noradrenergic agents into the ventricular system of the brain cause a release of vasopressin (46-48). Second, electrical stimulation of the A1 area in the medulla oblongata of the rat causes an increase in neural activity of the phasicaly firing vasopressin neurons (and no effect on oxytocin cells) which is eliminated by local injections of the noradrenaline neurotoxin 6-hydroxydopamine into the PVN (49). Lesions of either the dorsal or ventral noradrenergic pathways cause a decrease in the release of AVP elicited by hemorrhage (50). Thus, the question of whether noradrenaline is an excitatory or an inhibitory transmitter is unsettled. Moreover, it is possible that noradrenaline acts as a neuromodulator in this system, as it has been found in the cerebellum (51), enhancing either ongoing excitatory or inhibitory activity.

Another set of connections that may be involved in the control of vasopressin release is illustrated in Figure 6. Anatomical studies have indicated that the median preoptic nucleus which lies dorsal to the third ventricle provides an input to the SON (52). This nucleus has received a lot of attention because lesions of this area disrupt various forms of experimental hypertension (53). The major function of this area appears to be a site of integration of hormonal and neural inputs which regulate body fluid homeostasis. The median preoptic nucleus receives afferents from a number of structures that have been involved in cardiovascular control.

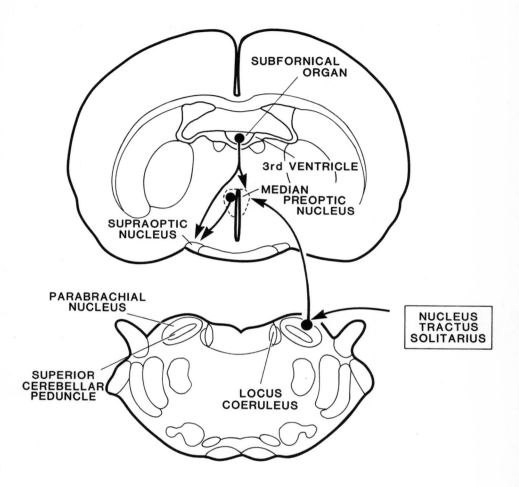

Figure 6: Inputs to the supraoptic nucleus come from the median preoptic nucleus, the subfornical organ, and parabrachial nucleus. The median preoptic nucleus receives inputs from parabrachial nucleus which is a major relay nucleus for cardiovascular information. The subfornical organ is likely to provide osmosensitive afferents to the supraoptic nucleus.

These include: 1) the subfornical organ (an circumventricular area containing angiotensin II neurons - 54); 2) the PVN; 3) the lateral parabrachial nucleus (23); 4) the nucleus tractus solitarius (55); and 5) the C1 adrenergic neurons (56). In addition, the SON and PVN receive inputs from angiotensin II cells of the subfornical organ (54); this is likely to be an osmosensitive neural pathway.

The subfornical neurons are thought to function as osmoreceptors and to be sensitive to circulating angiotensin II. Lesions of this region have been shown to attenuate AVP release induced by angiotensin II. Since there is a direct pathway from the subfornical organ to the SON and PVN, it is likely to be an important system involved in the control of vasopressin release.

REFERENCES

1. Malliani, A. Rev. Physiol. Biochem. Pharmacol. 44: 11-74, 1982.
2. Furness, J.B., Papka, R.E., Della, N.G., Costa, M., Eskay, R.L. Neuroscience 7: 447-459, 1982.
3. Foreman, R.D., Ohata, C.A. Am. J. Physiol. 238: H667-H674, 1980.
4. Ammons, W.S., Giradot, M.N., Foreman, R.D. J. Neurophysiol. 54: 73-81, 1985.
5. Panneton, W.M., Burton, H. Neuroscience 15: 779-797, 1985.
6. Saper, C.B. J. comp. Neurol. 210: 163-173, 1982.
7. Shipley, M.T. Brain Res. Bull. 8: 139-148, 1982.
8. Ciriello, J., Hrychshyn, A.W., Calaresu F.R. J. Auton. Nerv. Sys. 4: 43-61, 1981.
9. Wallach, J.H., Loewy, A.D. Brain Res. 188: 247-251, 1980.
10. Donoghue, S., Felder, R.B., Jordan, D., Spyer, K.M. J. Physiol., Lond. 347: 397-409, 1984.
11. Haeusler, G., Osterwald, R. Naunyn-Schmiedebergs Arch. exp. Path. Pharmak. 314: 111-121, 1980.
12. Gillis, R.A., Helke, C.J., Hamilton, B.L., Norman, W.P., Jacobowitz, D.M. Brain Res. 181: 476-481, 1980.
13. Furness, J.B., Elliot, J.M., Murphy, R., Costa, M., Chalmers, J.P. Neurosci. Lett. 32: 285-290, 1982.
14. Talman, W.T., Reis, D.J. Brain Res. 220: 402-407, 1981.
15. Dietrich, W.D., Lowry, O.H., Loewy, A.D. Brain Res. 237:

254-260, 1982.

16. Talman, W.T., Perrone, M.H., Reis, D.J. Science 209: 813-814, 1980.

17. Perrone, M.H. Brain Res. 230: 283-293, 1981.

18. Ricardo, J.A., Koh, E.T. Brain Res. 153: 1-26, 1978.

19. Loewy, A.D., Burton, H. J. comp. Neurol. 181: 421-450, 1978.

20. Leslie, R.A. Neurochem. Int. 7: 191-211, 1985.

21. Palkovits, M. Neurochem. Int. 7: 213-219, 1985.

22. Loewy, A.D., McKellar, S., Saper, C.B. Brain Res. 174: 309-314, 1978.

23. Saper, C.B., Loewy, A.D. Brain Res. 197: 291-317, 1980.

24. Helke, C.J., Neil, J.J., Massari, V.J., Loewy, A.D. Brain Res. 243: 147-152, 1982.

25. Sawchenko, P.E., Swanson, L.W. Brain Res. Rev. 4: 275-325, 1982.

26. Yamada, K.A., McAllen, R.M., Loewy, A.D. Brain Res. 297: 175-180, 1984.

27. Loewy, A.D., Sawyer, W.B. Brain Res. 245: 379-383, 1982.

28. Backman, S.B., Henry, J.L. Can. J. Physiol. Pharmacol. 62: 248-251, 1984.

29. Gilbey, M.P., McKenna, K.E., Schramm, L.P. Neurosci. Lett. 41: 157-159, 1983.

30. Cowley, A.W., Quillen, E.W., Skelton, M.M. Fed. Proc. 42: 3170-3176, 1983.

31. Harris, M.C. J. Endocr. 82: 115-125, 1979.

32. Kannan, H., Yagi, K. Brain Res. 145: 385-390, 1978.

33. Yamashita, H. Brain Res. 126: 551-556, 1977.

34. Bennett, T., Gardiner, S.M. Cardiovasc. Res. 19: 57-68, 1985.

35. Menninger, R.P. Fed. Proc. 44: 55-58, 1985.

36. Fuxe, K. Acta physiol. scand. 64: 37-85, supplement 247, 1965.

37. Lindvall, O., Bjorklund, A. Acta physiol. scand. 412: 1-48, 1974.

38. Ungerstedt, U. Acta physiol. scand. 367: 1-48, 1971.

39. Swanson, L.W., Sawchenko, P.E. Ann. Rev. Neurosc. 6: 269- 324, 1983.

40. Arnauld, E., Cirino, M., Layton, B.S., Renaud, L.P. Neuroendocrinology 36: 187-196, 1983.

41. Barker, J.L., Crayton, J.W., Nicoll, R.A. Science 171:

208-210, 1971.

42. Moss, R.L., Dyball, R.E.J., Cross, B.A. Brain Res. 35: 573-575, 1971.

43. Moss, R.L., Urban, I., Cross, B.A. Am. J. Physiol. 232: 310-318, 1972.

44. Armstrong, W.E., Sladek, C.D., Sladek, J.R. Endocrinology 111: 273-279, 1982.

45. Blessing, W.W., Sved, A.F., Reis, D.J. Science 217: 661-663, 1982.

46. Bhargava, K.P., Kulshrestha, V.K., Srivastava, Y.P. Br. J. Pharamacol. 44: 617-627, 1972.

47. Hoffman, W.E., Phillips, M.I., Schmid, P. Neuropharmacology 16: 563-569, 1977.

48. Kuhn, E.R. Neuroendocrinology 16: 255-264, 1974.

49. Day, T.A., Ferguson, A.V., Renaud, L.P. J. Physiol., Lond. 355: 237-249, 1984.

50. Everitt, B.J., Lightman, S.L., Todd, K. J. Physiol. 341: 81P, 1983.

51. Woodward, D.J., Moises, H.C., Waterhouse, B.D., Hoffer, B.J., Freedman, R. Fed. Proc. 38: 2109-2116, 1979.

52. Swanson, L.W. J. comp. Neurol. 167: 227-256, 1976.

53. Brody, M.J., Johnson, A.K. In: Disturbances in Neurogenic Control of Circulation, (Eds. F.M. Abboud et al.) pp. 105-117, 1981.

54. Lind, R.W., Swanson, L.W., Ganten, D. Brain Res. 321: 209-215, 1984.

55. Saper, C.B., Levisohn, D. Brain Res. 288: 21-31, 1983.

56. Saper, C.B., Reis, D.J., Joh, T. Neurosci. Lett. 42: 285-291, 1983.

2

BRAINSTEM MECHANISMS GOVERNING THE TONIC AND REFLEX CONTROL OF THE CIRCULATION

REIS, D.J., RUGGIERO, D.A., GRANATA, A.
Laboratory of Neurobiology, Cornell University Medical College, New York

INTRODUCTION

One of the classical problems of cardiovascular physiology has been the localization of neurons in the brainstem required for maintenance of normal levels of arterial pressure (AP). Experiments dating back to the mid-19th century demonstrate that removal of the brain rostral to the medulla oblongata had no effect on AP while transection of the spinal cord at the first cervical segment results in its collapse (1). Such studies demonstrated that neurons localized somewhere in the medulla provide sufficient excitation of preganglionic sympathetic neurons in the intermediolateral columns of the spinal cord to maintain vascular tone.

For many years the precise localization within the medulla of such neurons could never be established. Indeed, it was proposed that such neurons, rather than localized in a single site in the brainstem, were widely diffused throughout the rostral-caudal extent of the brain (2). However, over the past several years, evidence has accrued to indicate that such neurons are not distributed. Rather they reside in a highly restricted portion of the rostral ventrolateral medulla (RVL). We have proposed that within this zone the critical neurons are, or at the very least are in the immediate surround of a small population of neurons which synthesize adrenalin, the C1 group of Hökfelt et al. (3). We have termed this zone the C1 area (4,5). The evidence supporting our contention and the importance of neurons of the C1 area in cardiovascular control will be reviewed.

The C1 Area as the Tonic Vasomotor Center

Evidence implicating the general area of RVL as a cardiovascular control center has come from several lines of evidence. The first is pharmacological and based on findings that drugs applied to a restricted region of the ventral surface of the rostral medulla could lower AP to levels close to those produced by cord transection (6-10). Such agents include the inhibitory amino acids GABA and glycine, and also the drugs pentobarbitone, or clonidine. Since these agents act synaptically and not upon fibers of passage, the findings indicate that neurons and not projections through the RVL provide tonic excitatory input to the preganglionic sympathetic neurons. Lesions of the ventral medulla just beneath this region also replicated the effects of spinal cord transection upon AP (11,12). The demonstration of a direct projection from neurons in the subjacent medulla into segments of the thoracic cord harboring preganglionic neurons (13) provided an anatomical substrate for the control.

The second line of evidence came from more traditional studies of the effects of brain stimulation and lesions rather than studies directed at investigating the chemosensitive zones of the ventral medullary surface. Such studies (14-16) demonstrated that neurons lying within the RVL were sympathoexcitatory as were their purported pathways and that lesions of these neurons resulted in a fall of AP to spinal levels. Anatomical studies established that neurons in this area also projected into the spinal cord (14,15). The identity of the specific neuronal group within RVL responsible for the control of AP was not known.

Several years ago we were struck by the similarity between the localization within the RVL of the purported vasomotor neurons and the distribution of the C1 group of adrenalin synthesizing neurons first described by Hökfelt et al. (3). These neurons can be identified immunocytochemically by the presence of the adrenalin synthesizing enzyme phenylethanolamine N-methyltransferase (PNMT). We therefore undertook a series of studies to determine if the vasomotor neurons of the RVL corresponded to the C1 group.

We found, with careful mapping (Figure 1) (17,18) with antibodies to PNMT that the distribution of the C1 area of the rat corresponded almost exactly with the tonic vasomotor areas defined more generally in cat and rabbit in the RVL of cat and rabbit. Using tracer techniques combined with immunocytochemistry, we were able to establish that most of the neurons of the RVL which others had demonstrated to project to the spinal cord (13,15) contained PNMT and were, therefore, C1 neurons (18-21). Moreover, we were able to show using anterograde tracing methods that the spinal projection of these C1 cells was exclusively onto neurons of the intermediolateral

columns (20,21). This finding indicated that the function of C1 neurons in
the spinal cord was presumably exclusively to control autonomic function.

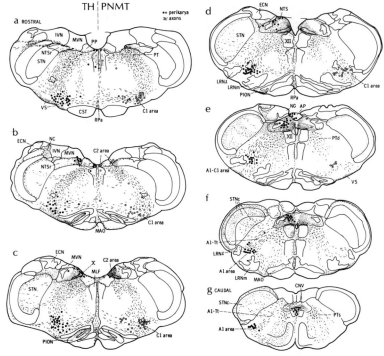

Fig. 1 Distribution of phenylethanolamine N-methyltransferase (PNMT, right
side) and tyrosine hydroxylase (TH, left side) immunoreactive
neurons and processes in rat medulla. Cells in rostral ventrolateral medulla (C1 area) synthesize epinephrine (A-C); those in
the caudal ventrolateral medulla synthesize norepinephrine (F-G)
and at intermediate levels both cell types are admixed (D-E). Axons
of C1 neurons arch dorsally to form a longitudinal fiber tract,
principal adrenergic tegmental bundle (PT). PT constitutes the
dorsal-most segment of the more extensive and heterogenous TH-
labeled fiber system. Fibers of PT ascend and descend. At caudal
levels (D-G) descending limb of PT (PTd), also referred to as
spinal limb (PTs), shifts medially and sprays ventrolaterally and
caudally (E-G) en route to the lateral funiculus of the spinal
cord. Note that A1 fibers do not contribute to PTd and rather pass
lateral to it toward NTS (D-G). PNMT-stained perikarya (open

circles); TH-stained perikarya (filled circles); PNMT- and TH-labeled axons (lines); PNMT- and TH-labeled terminals (stipple). A1-Tt, norepinephrine transtegmental tract; A1, ventrolateral norepinephrine cell area; AP, area postrema; C1, ventrolateral epinephrine cell area; C2, dorsomedial epinephrine cell area; CNV, commissural nucleus of vagus; CST, corticospinal tract; ECN, external cuneate nucleus; IVN, inferior vestibular nucleus; LRN1, precerebellar lateral reticular nucleus pars medialis; MAO, medial accessory olive; MLF, medial longitudinal fasciculus; MVN, medial vestibular nucleus; NC, nucleus cuneatus; NG, nucleus gracilis; NTSr, nucleus tractus solitarii pars rostralis; PION, principal nucleus of inferior olive; PP, nucleus prepositus; RPa, nucleus raphe pallidus; STN, spinal trigeminal nucleus; STNc, caudal sub-division of STN; VS, ventral medullary surface; X, dorsal vagal motor nucleus; XII, hypoglossal nucleus (from Ref. 24).

The region of the RVL from which electrical excitation elevated AP corresponds exactly to the C1 area (Figure 2) (20, 22). Electrical stimulation of the pathway within the medulla through which the PNMT containing fibers coursed also elicits changes in circulation (22) comparable to those produced by stimulation of the cells of origin in the C1 area. Thus, the vasopressor zones of the medulla and the C1 neuronal system overlap.

In contrast to the effects of stimulation, bilateral electrolytic lesions of the C1 area (22,24), application of GABA or glycine to the area (22) or inactivation of the region with tetrodotoxin (Figure 3) (22, 25) lowers AP to spinal levels. Stimulation or lesions of neurons in the immediate surround do not elicit comparable results.

Recently, we were able to demonstrate (11) that the C1 area lies just over those regions of the ventral surface of the medulla from which chemical stimulation with glutamic acid will elevate (Figure 4) while glycine or GABA will lower AP. In a series of studies we have been able to show that the effects elicited by chemical stimulation of the ventral surface of the medulla are mediated by C1 neurons (11), and that the anatomical basis of the action of agents on the ventral surface is probably to stimulate processes of C1 neurons that lie in immediate apposition to the ependymal surface (11,18). Thus, the responses elicited from the

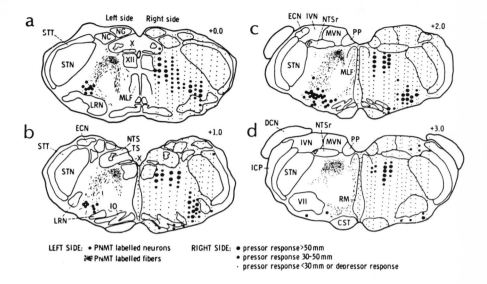

Fig. 2 Locations of the most active medullary pressor regions to constant
current electrical stimulation compared to locations of C1 cells
and fibers immunocytochemically labeled for PNMT. Left side of each
section, PNMT-labeled C1 neurons and PNMT-labeled fibers. Right side
of each section, pressor responses to electrical stimulation with 25
uA (100 Hz, 0.5 msec, 10-sec train). Responses between 30 and 50 mm
Hg are indicated by small solid circles, whereas those greater than
50 mm Hg are indicated by larger solid circles. Responses less than
30 mm Hg or depressor responses are indicated by small dots. The
location of the pressor region in the ventrolateral medulla corresponds to the location of the C1 neurons, and the location of the
pressor region in the dorsomedial medulla corresponds to the
location of the PNMT-labeled fiber bundle. CST, corticospinal
tract; DCN, dorsal cochlear nucleus; ECN, external cuneate nucleus;
ICP, inferior cerebellar peduncle; IO, inferior olive; IVN,
inferior vestibular nucleus; LRN, lateral reticular nucleus; MLF,
medial longitudinal fasciculus; MVN, medial vestibular nucleus; NC,
nucleus cuneatus; NG, nucleus gracilis; NTS, nucleus tractus
solitarius; NTSr, nucleus tractus solitarius pars rostralis; PP,

nucleus prepositus; RM, raphe magnus; STN, spinal trigeminal nucleus; STT, spinal trigeminal tract; TS, tractus solitarius; VII, facial nucleus; X, dorsal motor nucleus of vagus; XII, hypoglossal nucleus (from Ref. 22).

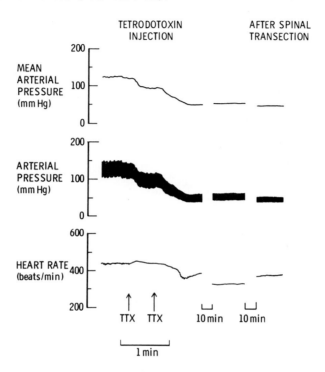

Fig. 3 Cardiovascular responses to bilateral injections of tetrodotoxin (10 pmole/100 nl saline) into the C1 area. The first injection (TTX), indicated by an arrow on the bottom of chart recording, produced a partial reduction of AP, while injection into the contralateral side (second arrow) resulted in a collapse of AP to levels comparable to that produced by subsequent spinal cord transection. CST = corticospinal tract; IO = inferior olive; IVN = inferior vestibular nucleus; MVN = medial vestibular nucleus; NA = nucleus ambiguus; NTSr = nucleus tractus solitarius pars rostralis; PP = nucleus prepositus; RPa = raphe pallidus; STN = spinal trigeminal nucleus; STT = spinal trigeminal tract; TS = tractus solitarius (from Ref. 22).

ventral medullary surface and those elicited from within the brain act through a common neuronal pool, the C1 cells.

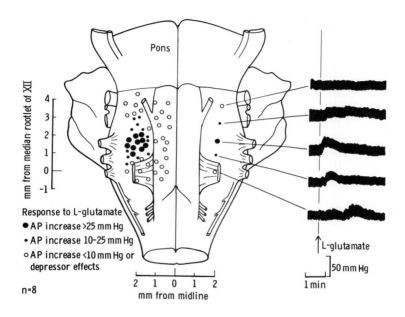

Fig. 4 Map of the ventral surface of the rat medulla showing evoked changes in arterial pressure (AP: mmHg) to topical unilateral application of L-glutamate (5 nmole). The left side of the figure shows the results of 8 experiments (circles represent the points of application of the pipette containing the drug). Records of AP obtained in a typical case are depicted on the right side. The arrow indicates the time of onset of application of the drug (from Ref. 10).

Our studies, therefore, established that C1 neurons project into the intermediolateral columns of the spinal cord, are sympathoexcitatory and probably are tonically active. The identity of the neurotransmitter released from the endings of C1 neurons in the spinal cord to produce sympathoexcitation, however, is unknown. Indeed, one of the problems still to be clarified is highlighted by the paradox that pharmacological evidence suggests that epinephrine inhibits rather than excites preganglionic spinal neurons (see 24,25 for discussion). However, since the C1 neurons contain other putative transmitters including NPY (26), substance P (27) and possibly other biologically active agents, it is entirely conceivable that a substance co-stored by PNMT cells may be the excitatory agent acting upon spinal preganglionic neurons. Thus, while the adrenalin neurons of the C1 group appear to be the tonic vasomotor neurons, adrenalin may not be the neurotransmitter which elicits sympathoexcitation.

Role of C1 Area Neurons and Baroreceptor Reflexes

The identification of the region and the most probable cellular elements that serve as tonic vasomotor neurons have led to further characterization of the role of the C1 area in physiological integration of the circulation. Our investigations, so far, have been directed to establishing the possible role of these cells in mediating other cardiovascular reflex events and to determine the identity of neurotransmitters which may be acting to modulate the discharge of neurons in this region.

One critical function of the C1 area neurons is in mediating the vasodepressor response elicited from stimulation of arterial baro- and other cardiopulmonary receptors whose afferent fibers are carried in the IXth and Xth cranial nerves (28-30). The anatomical substrate of this control is via the direct projection of these visceral afferents to regions of the nucleus tractus solitarii (NTS) within which the afferent fibers of the IXth and Xth nerves terminate (31). These regions of NTS are essential for integration of all cardiopulmonary reflexes.

We have found that interference with function of C1 neurons will abolish the fall of AP elicited by stretch of arterial baroreceptors in the carotid sinus or electrical stimulation of the vagus nerves (Figure 5) (23,24). It also abolishes the elevation of AP produced by disruption of baroreceptor reflexes by lesions in the NTS (32).

In support of the view that the NTS-C1 pathway mediates baroreceptor reflexes, are recent electrophysiological studies of single neurons contained within the C1 area which project monosynaptically into the spinal cord (33,34). Many neurons in this region discharge with a rhythmic pattern of a similar period to that of the heart rate. This rhythmic pattern is driven by baroreceptor afferents. The findings, therfore, suggest that C1 neurons are tonically inhibited by an input from barorecptor activity through neurons in the NTS and that the fall of AP resulting from transient increases in baroreceptor actively or, conversely, the transient elevations elicited by withdrawal of such drive, depends upon modulation of the discharge of C1 neurons, in turn, determining the output of preganglionic sympathetic neurons.

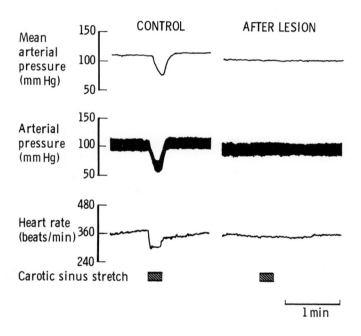

Fig. 5 Effect of lesion of left C1 area in anesthetized rat in which the right NTS was lesioned. Lesion of the right NTS restricts the baroreceptor loop to an ipsilateral pathway from baroreceptors to NTS to C1 area. This experiment conclusively demonstrates that C1 neurons mediate the vasodepression response from baroreceptor stimulation (from Ref. 24).

GABAergic Inhibition of C1 Neurons by Local Neurons

We have recently identified two other inhibitory influences on the tonic vasomotor outflow of the C1 area. One is mediated by the release of GABA. It has long been known that GABA applied either to the ventral surface of the medulla or directly within the C1 area will lower AP (11,22,34-37). Thus, GABA receptors lying upon neurons in close apposition to or upon C1 neurons will, when stimulated, inhibit outflow to preganglionic neurons and hence lower AP (Figure 6). More important are the findings that local application of the GABA antagonist bicuculline to the C1 area will elevate AP (Figure 6) (22). This latter observation indicates that there is a tonic inhibitory control of AP from the release of GABA by nerve terminals in the C1 area.

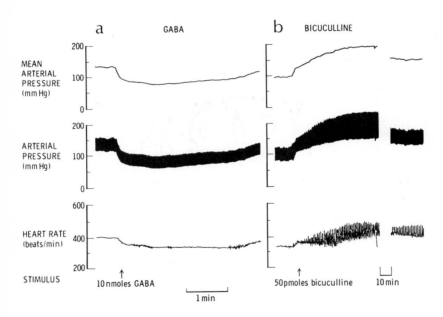

Fig. 6 Cardiovascular responses to unilateral injections into the RVL of (a) GABA (10 nmol) and (b) bicuculline (50 pmol). GABA caused a decrease in AP and heart rate, whereas the specific GABA antagonist bicuculline caused a long-lasting elevation of AP and tachycardia (from Ref. 22).

Recent findings have suggested that the source of much, if not all, of the GABA within the C1 area is from an abundant network of local GABAergic neurons. These neurons have been identified in the C1 area immunocytochemically by staining with antibodies to the GABA synthesizing enzyme glutamic acid decarboxylase (GAD) (38,39). GABAergic neurons of the C1 area have been recently demonstrated by EM immunocytochemistry to directly contact PNMT-containing cells in the region (40). The evidence suggests, therefore, that most of the GABAergic input onto C1 cells is via local interneurons. The fact that bicuculline elevates AP to levels comparable to that produced by maximal electrical stimulation suggests that such inhibition is potent and raises the interesting question as to the pathways and physiological relevance of this purportedly interneuronal control of the discharge of C1 cells. Conceivably, it may participate in mediating the baroreceptor reflex (41), although it seems unlikely that this control is mediated via a direct GABAergic projection from the NTS to the C1 area (38).

Role of the Caudal Ventrolateral Medulla and Noradrenergic Control of C1 Discharge

A second inhibitory input arises from the neurons of the caudal ventrolateral medulla largely in the surround of a region containing noradrenergic neurons of the A1 group (25,43-46). Neurons in this region have cardiovascular activities entirely opposite from those of neurons in the C1 area. When excited electrically or chemically by glutamate or the glutamate analogue kainic acid, they elicit a reduction of sympathetic activity and a reduction of AP (Figure 7) (25,43,45,46). In contrast, inactivation of these neurons by electrolytic lesions, by application of large doses of the glutamate analogue kainic acid or by GABA increases sympathetic discharge and AP (Figure 6) (42,44-46). Indeed, the inactivation of neurons in this region results in fulminating hypertension in part the consequence of sympathetic nerve activity, although the release of arginine vasopressin from the posterior pituitary contributes (44,46).

That the C1 area is critical for the expression of the hypertension elicited by inactivation of A1 neurons has been demonstrated by abolishing all responses from the A1 area and by inactivating the C1 area with tetrodotoxin (Figure 6). Thus, neurons in the caudal medulla projecting to the C1 area exert some tonic inhibition of sympathetic activity. The identity of the neurons in the caudal medulla which mediate this inhibition is unclear but there is suggestion that the noradrenergic neurons may play a part.

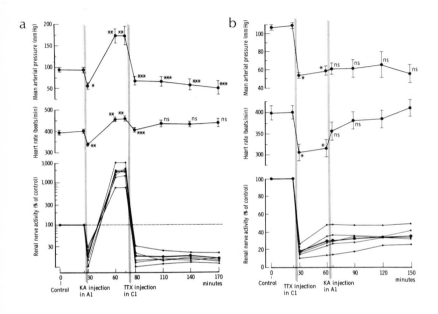

Fig. 7 Effect of bilateral microinjection of kainic acid in the caudal ventrolateral medulla on AP, heart rate and renal nerve discharge. Values ± SEM from 6 experiments. Kainic acid 400 pmole in 0.1 ul of saline. (a) Control. Note initial fall of AP followed by rise representing the initial stimulation and delayed inactivation of neurons. (b) All responses are blocked by bilateral injection of tetrodotoxin CMX, 10 pmole in 0.1 ul saline (from Ref. 25).

That the release of noradrenalin in the C1 area is sympathoinhibitory can clearly be shown by the demonstration that the local application of adrenergic agonists such as alpha-methyl norepinephrine, or the indirect agonist tyramine will lower blood pressure in a dose dependent way (Figure 8) (25). The effects of tyramine on lowering arterial pressure can be abolished by the pretreatment of the region with a local injection of reserpine or by destruction of the local catecholaminergic nerve terminals by 6-hydoxydopamine (25). Since the C1 area, however, may receive inputs from other catecholaminergic projections including those of the A5 and A2 groups (see 25 for discussion), it is still conceivable that some of the noradrenergic release is from these groups. However, destruction of both the A2 and A5 groups does not block the hypertension elicited by lesions of the caudal medulla, strongly suggesting that indeed the source of the noradrenalin lowering AP is from the A1 group.

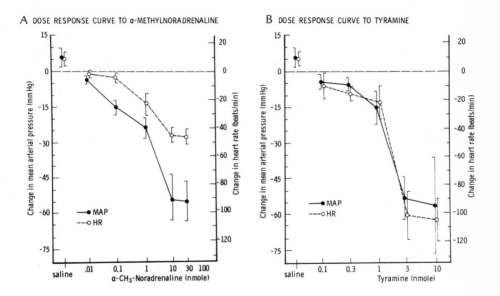

Fig. 8 Dose dependent hypotension and bradycardia elicited by micro-injection (0.1 ul in saline) into the C1 area of (a) alpha-methyl noradrenaline or (b) tyramine. Each dose represents mean ± SEM from 4-8 rats (from Ref. 25).

The activation of alpha adrenergic receptors in the C1 area, however, may have important therapeutic implications for hypertension. It is well known that the RVL is a site at which clonidine will act to lower AP (25,47) and contains abundant numbers of alpha-2 adrenergic receptors (48-50). Privitera, et al. (51) have recently demonstrated that the local microinjections of beta blocking agents such as propranolol into the C1 area will lower AP presumably by indirect action of releasing norepinephrine. These findings, therefore, raise the prospect that a number of drugs efficacious in the treatment of hypertension in man may exert much of their action to lower AP through the C1 area and gives further strength to the view that C1 area neurons not only are important in the normal integration of cardiovascular function but may play a role in abnormal states of AP control.

There is evidence now that acetylcholine may be an excitatory agent within the C1 area. The region contains muscarinic receptors (52), there are large numbers of neurons in the surround of the C1 area containing the acetylcholine synthesizing enzyme choline acetyltranferase (CAT) (52), the local application of cholinergic agonists elevates AP via muscarinic receptors (53,54), and the application of cholinergic antagonists in the region will lower AP. Since neurons in the C1 area are tonically active, the question, of course, arises as to the driving excitatory background for this activity. One possiblity is that cholinergic neurons serve as pacesetters for the process.

SUMMARY AND CONCLUSION

A region of the brainstem has now been identified which, all evidence suggests, contains the long sought for neurons essential for providing excitatory background tone to preganglionic sympathetic neurons and functioning as the tonic vasomotor center. These neurons are congregated within a small area of the rostral ventrolateral medulla containing adrenergic neurons of the C1 group (the C1 area). It is our contention that in all probability, the C1 neurons themselves are the critical elements. The C1 area neurons project directly into intermediolateral columns of the cord, and are tonically active.

A potent inhibition is exerted upon C1 area neurons by pathways mediating baroreceptor responses emanating from the NTS. In this manner, C1 neurons are the critical link through which baroreceptor and other information arising from arterial baroreceptors lowers AP. This fact has permitted us to chart a provisional map of the baroreceptor reflex pathway (Figure 9). Neurons of the C1 area are also probably of critical importance in hypertension in that they are essential for the expression of the elevation of AP associated with at least one mode of experimental hypertension, that produced by interference of baroreceptor afferent activity. Moreover, they appear to be a critical site for the antihypertensive actions of a number of drugs of clinical importance in lowering AP, including clonidine and possibly the beta blockers. The identification of a particular cellular population of such importance in the normal and abnormal control of AP provides a basis for further investigations into the cellular and molecular mechanisms governing circulatory control in health and disease (5).

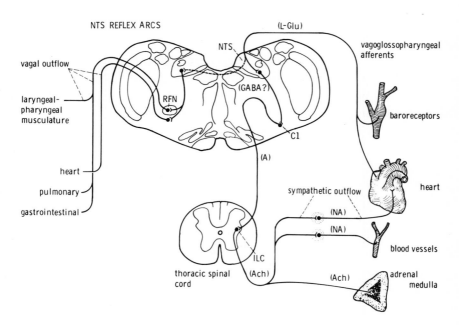

Fig. 9 Pathway of baroreceptor and other vasodepressor pathways. Abbreviations as in Figure 1 except A, adrenaline; ACh, acetylcholine; NA, noradrenaline (from Ref. 31).

ACKNOWLEDGEMENTS

This research was supported by grants from NIH and DOD.

REFERENCES

1. Alexander, R.S. J. Neurophysiol. 9: 205-217, 1946.
2. Hilton, S.M. Brain Res. 87: 213-219, 1975.
3. Hökfelt, T., Fuxe, K., Goldstein, M., Johansson, O. Brain Res. 66: 235-251, 1974.
4. Reis, D.J. Circulation 70: (Suppl. III) III-31-45, 1984.
5. Reis, D.J. J. Cardiovasc. Pharmacol. 7: (Suppl. 2) S160-166, 1985.
6. Feldberg, W., Guertzenstein, P.G. J. Physiol. (Lond.) 224: 83-103, 1972.
7. Guertzenstein, P.G. J. Physiol (Lond.) 229: L395-408, 1973,
8. Bousquet, P., Feldman, J., Velly, J., Bloch, R. Eur. J. Pharmacol. 34: 151-156, 1975.
9. Feldberg, W., Guertzenstein, P.G. J. Physiol. 229: 395-408, 1976.
10. Benarroch, E.E., Granata, A.R., Ruggiero, D., Park, D.H., Reis, D.J. Am. J. Physiol. 1986 (in press).
11. Guertzenstein, P.G., Silver, A. J. Physiol. (Lond.) 242: 489-503, 1974.
12. Guertzenstein, P.G., Hilton, S.M., Marshall, J.M., Timms, R.J. J. Physiol. (Lond.) 275: 78-79, 1978.
13. Amendt, K., Czachurski, J., Dembowsky, K., Seller, H. Pflü. Arch. 375: 289-292, 1978.
14. Dampney, R.A.L., Moon, E.A. Am. J. Physiol. 239 (Heart Cir. Physiol. 8): H349-H358, 1980.
15. Dampney, R.A.L., Goodchild, A.K., Robertson, L.G., Montgomery, W. Brain Res. 249: 223-235, 1982.
16. Kumada, M., Dampney, R.A.L., Reis, D.J. Circ. Res. 45: 63-70, 1979.
17. Armstrong, D.M., Ross, C.A., Pickel, V.M., Joh, T.H., Reis, D.J. J. Comp. Neurol. 212: 173-187, 1982.
18. Ruggiero, D.A., Ross, C.A., Anwar, M., Park, D.H., Joh, T.H., Reis, D.J. J. Comp. Neurol. 239: 127-154, 1985.
19. Ross, D.A., Armstrong, D.M., Ruggiero, D.A., Pickel, V.M., Joh, T.H. Reis, D.J. Neurosci. Lett. 25: 257-262, 1981.
20. Ross, C.A., Ruggiero, D.A., Joh, T.H., Park, D.H., Reis, D.J. Brain Res. 273: 356-361, 1983.
21. Ross, C.A., Ruggiero, D.A., Joh, T.H., Park, D.H., Reis, D.J. J. Comp. Neurol. 228: 168-184, 1984.

22. Ross, C.A., Ruggiero, D.A., Park, D.H., Joh, T.H., Sved, A.F., Fernandez-Pardal, J., Saavedra, J.M., Reis, D.J. J. Neurosci. 4: 479-494, 1984.

23. Granata, A.R., Ruggiero, D.A., Park, D.H., Joh, T.H., Reis, D.J. Hypertension 5 (Suppl.): 80-84, 1983.

24. Granata, A.R., Ruggiero, D.A., Park, D.H., Joh, T.H., Reis, D.J. Am. J. Physiol. (Heart 17) 248: H547-H567, 1985.

25. Granata, A.R., Numao, Y., Kumada, M., Reis, D.J. Brain Res. 1986 (in press).

26. Hökfelt, T., Lundberg, J.M., Tatemoto, K., Mutt, V., Terenius, J., Polak, S., Bloom, S., Sasek, C., Elde, R., Goldstein, M. Acta Physiol. Scand. 117: 315-318, 1983.

27. Lorenz, R.G., Saper, C.B., Wong, D.L., Ciaranello, R.D., Loewy, A.D. Neurosci. Lett. 55: 255-260, 1985.

28. Miura, M., Reis, D.J. Am. J. Physiol. 217: 142-153, 1969.

29. Panneton, W.M., Loewy, A.D. Brain Res. 191: 239-244, 1980.

30. Kalia, M., Sullivan, J.M. J. Comp. Neurol. 211: 248-264, 1982.

31. Ross, C.A., Ruggiero, D.A., Reis, D.J. J. Comp. Neurol. 242: 511-534, 1986.

32. Benarroch, E.E., Granata, A.R., Giuliano, R., Reis, D.J. Hypertension 1986 (in press).

33. Brown, D.L., Guyenet, P.G. Am. J. Physiol. 247 (Regulatory Integrative Comp. Physiol. 16): R1009-R1016, 1984.

34. Morrison, S., Milner, T., Pickel, V., Reis, D.J. Fed. Proc. 1986 (in press).

35. Yamada, K.A., Norman, W.P., Hamosh, P.P., Gillis, R.A. Brain Res. 248: 71-78, 1982.

36. Willette, R.M., Krieger, A.J., Barcas, P.O., Sapru, H.N. J. Pharmacol. Exp. Ther. 226: 893-899, 1983.

37. Keeler, J.R., Shults, C.W., Chase, T.H., Helke, C.J. Brain Res. 297: 217-224, 1984.

38. Meeley, M.P., Ruggiero, D.A., Ishitsuka, T., Reis, D.J. Neurosci. Lett. 58: 83-89, 1985.

39. Ruggiero, D.A., Meeley, M.P., Anwar, M., Reis, D.J. Brain Res. 339: 171-177, 1985.

40. Milner, T.A., Chan, J., Massari, V.J., Oertel, W.H., Park, D.H., Joh, T.H., Reis, D.J., Pickel, V.M. Abst. Soc. Neurosci. 11: 570, 1985.

41. Yamada, K.A., McAllen, R.M., Loewy, A.D. Brain Res. <u>297</u>: 175-180, 1984.

42. Blessing, W.W., West, M.J., Chalmers, J. Circ. Res. <u>49</u>: 949-958, 1981.

43. Blessing, W.W., Reis, D.J. Brain Res. <u>253</u>: 161-171, 1982.

44. Blessing, W.W., Sved, A.F., Reis, D.J. Science <u>217</u>: 661-662, 1982.

45. Blessing, W.W., Reis, D.J. Neurosci. Lett. <u>37</u>: 57-62, 1983.

46. Imaizumi, T., Granata, A.R., Benarroch, E.E., Sved, A.F., Reis, D.J. J. Hypertens. <u>3</u>: 491-501, 1985.

47. Bousquet, P., Schwartz, J. Biochem, Pharmacol. <u>32</u>: 1459-1465, 1983.

48. Unnerstall, J.R., Kopajtic, T.A., Kuhar, M.H. Brain Res. Rev. <u>319</u>: 69-101, 1984.

49. Ernsberger, P., Meeley, M.P., Reis, D.J. Abst. Soc. Neurosci. <u>11</u>: 490, 1985.

50. Ernsberger, P.R., Mann, J.J., Reis, D.J. Fed. Proc. 1986 (in press).

51. Privitera, P.J., Granata, A.R., Reis, D.J., Tackett, R.L., Gaffney, T.E. Abst. Soc. Neurosci. <u>11</u>: 192, 1985.

52. Arneric, S.P., Giuliano, R., Ernsberger, P., Underwood, M.D., Reis, D.J. Fed. Proc. 1986 (in press).

53. Willette, R.N., Punnen, S., Krieger, A.J., Sapru, H.N. J. Pharmacol. Exp. Ther. <u>231</u>(2): 457-463, 1984.

54. Benarroch, E.E., Guiliano, R., Ernsberger, P., Reis, D.J. Abst. Soc. Neurosci. <u>11</u>: 491, 1985.

3

ROSTRAL SOLITARY TRACT LESIONS PRODUCE VASOPRESSIN DEPENDENT HYPERTENSION IN THE DOG

BARNES, K.L., AVERILL, D.B., CONOMY, J.P., FERRARIO, C.M.

Research Division, Cleveland Clinic Foundation, Cleveland, Ohio

ABSTRACT

A possible reason for inconsistent results in attempts to produce NTS lesion-dependent hypertension in larger animals such as the dog may be incomplete destruction of baroreceptor afferents. Neuroanatomical studies in the dog revealed that the 8 mm length of the distribution of baroreceptor afferents within the nucleus tractus solitarius makes interruption of the baroreflex pathway by lesioning this nucleus unfeasible. However, the discovery that the baroreceptor afferents are bundled together in the solitary tract between 5 mm and 7 mm anterior to the obex suggested that bilateral destruction of the tracts would produce central baroreflex interruption. Pharmacological evaluations in dogs following rostral solitary tract lesions revealed both successful baroreflex interruption and an acute hypertension which was primarily dependent on increased circulating vasopressin. However, the similar decrease in arterial pressure following ganglionic blockade in both solitary tract lesioned and sham lesioned animals leaves the importance of augmented sympathetic outflow in solitary tract lesion hypertension in doubt.

Although the intermediate portion of the nucleus tractus solitarii (NTS) has long been known as the site of the first relay for the baroreceptor afferents (1) in the central nervous system (CNS), a number of investigators have encountered difficulty in producing hypertension that is sustained for weeks or months after central interruption of this CNS input. Doba and Reis (2) first showed that bilateral NTS lesions in the rat produced severe hypertension leading to acute cardiac failure and pulmonary edema. In contrast, Buchholz and Nathan (3) found that somewhat more restricted NTS lesions which abolished the reflex bradycardia to the rise in blood pressure following phenylephrine injections produced only lability of arterial pressure without altering its average level in rats studied between one week and 7 months after operation. Similarly, the hypertension which followed NTS lesions in the cat was reported to be moderate and associated with lability (4). Previous studies in the dog have yielded conflicting results. Carey et al. (5) reported mild hypertension during ten days of observation, Laubie and Schmitt (6) described severe hypertension sustained for several months after NTS lesions, whereas we found mild hypertension with lability during one to two months of post-lesion study (7). Although these inconsistencies in NTS lesion dependent hypertension may be attributed to differences in lesioning and/or recording techniques, the controversy is far from settled. It is possible that hypertension of neurogenic origin is associated with only intermittent bouts of blood pressure elevation. In the final analysis either labile or sustained hypertension may both have the same deleterious consequences upon the cardiovascular system if they occur over a period of time measured in years rather than weeks or months.

NEUROANATOMICAL LOCALIZATION OF THE BARORECEPTOR AFFERENT PATHWAYS IN THE DOG'S MEDULLA

A possible explanation for the failure to produce hypertension by central baroreceptor deafferentation in species other than the rat may be incomplete destruction of baroreceptor afferents. Studies in cat (8,9), rat (10) and rabbit (11) have revealed that the vagal, aortic depressor and carotid sinus afferent fibers have widespread terminations in the NTS. These observations would suggest that achieving complete and selective central baroreceptor deafferentation by lesioning the NTS would be very difficult, particularly in larger animals such as the dog and cat. Therefore we undertook a neuroanatomical investigation of the pathways in the dog brain stem traversed by the vagal (12) and carotid sinus nerve (13) afferents in order to find a site where a discrete lesion would produce complete interruption of the baroreceptor input to the CNS.

Horseradish peroxidase (HRP) was microinjected into either the nodose ganglion or the carotid sinus nerve. The brain stems were perfused, and serial sections processed for HRP histochemistry (14). As shown in Figure 1, vagal afferent fibers enter to dorsolateral medulla in a single bundle about 7 mm anterior to the obex. Since microinjection of HRP into the vagal afferent fibers is likely to label baroreceptor afferents from the aortic nerve, the vagal afferent projections can be expected to include the trajectory of these aortic baroreceptor fibers. The afferents travel dorsomedially to the solitary tract (TS), where they turn caudally, and remain in the TS until about 4.5 mm anterior to the obex, where they begin to enter the medial and lateral NTS, dorsal motor nucleus of the vagus and area postrema. Fibers and terminals can be seen in the NTS as far caudally as 3.5 mm behind the obex (13).

The carotid sinus afferents were seen to enter the medulla rostral to the vagal afferents, about 9 mm anterior to the obex (Figure 1). These fibers travel medially to the rostral solitary tract, where they turn caudally. The carotid sinus afferents also remain in the TS until 5 mm anterior to the obex, where they begin to distribute into the subnuclei of the NTS and the area postrema. Labelled fibers and terminals are found in the NTS caudally to about 3 mm behind the obex.

These neuroanatomical studies provided us with two important findings. First, the 8 mm length of the distribution of the carotid sinus and vagal afferent fibers and terminations within the NTS clearly makes destruction of the baroreceptor afferent relay by lesioning this nucleus unfeasible. However, the observation that the vagal and carotid sinus afferents are tightly bundled together in the solitary tract between 5 and 7 mm anterior to the obex suggested that discrete bilateral destruction of the TS at this level might be a means to produce central interruption of the baroreceptor afferents.

THE DEVELOPMENT OF ACUTE HYPERTENSION AFTER ROSTRAL SOLITARY TRACT LESIONS IN THE DOG

In order to locate the rostral solitary tracts containing the baroreceptor afferents for subsequent lesions, we developed a technique for monitoring evoked potentials from the dorsal medulla about 6 mm anterior to the obex during electrical stimulation of the cervical vagus nerve (15). In 14 mongrel dogs (12 $\pm$ 1 kg body weight) anesthetized with halothane, stimulating electrodes embedded in silastic cuffs were placed around both vagosympathetic nerve trunks in the neck, and catheters were inserted into a femoral artery and vein for measurement of aortic blood pressure and administration of drugs. After positioning the animal in a stereotaxic frame angled at 50 ° (nose down) and exposing the dorsal medulla, computer-averaged evoked potentials were recorded from the brain stem with low impedance microelectrodes (F. Haer Co., Brunswick, ME) during monophasic electrical stimulation (0.5-1.5 V, 30 Hz, 0.2 msec pulse duration) of the ipsilateral vagosympathetic trunk. Systematic exploration of the dorsal medulla between 5 and 7 mm anterior to obex revealed that the maximum vagal evoked potential was obtained when the electrode tip was in the TS. In nine dogs the solitary tracts were lesioned bilaterally (2-5 mA d.c., 30 sec) with a larger monopolar electrode (Rhodes NEX-100, Woodland Hills, CA). In five other animals the solitary tracts were located by finding the maximal vagal evoked potential, but not lesioned, thus providing a sham lesioned group. After completion of the surgical procedure including discontinuation of anesthesia and codeine (15 mg i.m.), the animal was transferred to a quiet, darkened laboratory. Arterial blood pressure and heart rate were recorded continuously for the remainder of the experiment; the recorded signals were processed by a digital computer, as described previously (16).

The development of hypertension was monitored for two and one half hours after lesion placement. During this period, the baroreflex control of heart rate was evaluated by determining the interbeat interval during the rise in arterial pressure produced by phenylephrine (Neo-Synephrine , 10 ug/kg i.v., Winthrop Laboratories, New York, NY). We then examined the relative contributions of arginine vasopressin (AVP) and the sympathetic nervous system to the level of blood pressure in both TS lesioned and sham lesioned animals. The contribution of vasopressin to the level of arterial blood pressure was assessed by administering an antagonist of the vascular actions of AVP (d(CH$_2$)$_5$Tyr (Me) AVP, 20 ug/kg i.v.) two and one half hours after TS lesion or sham lesion. The participation of the sympathetic nervous system in the level of arterial pressure was determined by ganglionic blockade with hexamethonium chloride (15 mg/kg i.v., U.S. Biochemical Corp., Cleveland, OH) given three hours after TS lesion or sham lesion.

At the completion of the experiment the brain stem was removed and placed in a mixture of 10 % formalin and 30 % sucrose for at least two weeks. Serial sections were cut at 50 um, stained with neutral red and Luxol fast blue, and examined to determine the rostrocaudal extent of bilateral interruption of the solitary tracts. All results are expressed as mean ± SEM. Analysis of variance (17) was applied to the data, and Dunnett's test (18) was used to evaluate differences of successive measurements from the control condition. Differences were considered significant for $p < 0.05$.

In all of the nine TS lesioned dogs interruption of the baroreflex pathway was documented by both the absence of bradycardia in response to i.v. phenylephrine and histological evidence of bilateral destruction of the solitary tracts between 5 and 7 mm anterior to the obex. In contrast, all five sham lesioned animals demonstrated baroreflex control of heart rate.

Mean arterial pressure (MAP) rose rapidly after anesthesia was discontinued in the TS lesioned dogs, but remained stable in the sham lesioned animals. Ninety minutes after placement of the second TS lesion MAP was 164 ± 5 mm Hg in the TS lesioned group, more than 40 mm Hg greater than in the sham lesioned animals (120 ± 3 mm Hg, $p < 0.05$). However, heart rates in the two groups of dogs were not different. Figure 2 shows that administration of the AVP antagonist (20 ug/kg i.v.) to the TS lesioned animals two and one half hours after the lesions decreased MAP by an average of 34 mm Hg, from 157 ± 5 mm Hg to 123 ± 8 mm Hg ($p < 0.01$), essentially the same level as the sham lesioned dogs. This decrease in MAP was sustained for more than 30 minutes. In contrast, MAP was not affected by the vasopressin antagonist in the sham lesioned dogs (-1 ± 2 mm Hg). Three hours following surgery ganglionic blockade with hexamethonium chloride (15 mg/kg i.v.) reduced mean arterial pressure by amounts that did not differ significantly (TS lesioned: - 86 ± 6 mm Hg; sham lesioned: -64 ± 4 mm Hg), with MAP plateauing at 48 ± 4 mm Hg in the TS lesioned dogs and at 58 ± 5 mm Hg in the sham lesioned group. These data indicate that in the presence of an augmented sympathetic outflow an increase in plasma vasopressin makes a significant contribution to the acute hypertension produced by bilateral TS lesions.

DISCUSSION

The studies detailed above demonstrate how neuroanatomical information

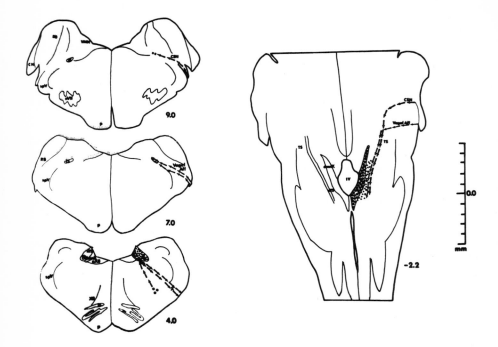

Figure 1: Pathways in the dog's medulla traversed by the vagal and carotid
sinus nerve afferent fibers. Left side: transverse sections of
the dog's medulla at 9, 7, and 4 mm anterior to the obex. Dashed
lines on the left side of each section indicate the path of the
carotid sinus (9 mm) and vagal (7 mm) afferents, as well as the
rostral most vagal efferent fibers (4 mm). Right side: horizontal section of the dog's medulla at the level of the dorsal motor
nucleus of the vagus. Dashed lines indicate the trajectories of
the vagal and carotid sinus nerve afferent fibers entering the
left side of the medulla and running caudally in the solitary
tract. Small dots medial to the tract indicate the distribution
of the afferents in the nucleus tractus solitarii at this level.
Larger dots in the dorsal motor nucleus of the vagus represent
vagal motor neurons. Scale indicates the level of the obex at
0.0. CN = cochlear nucleus, CSN = carotid sinus nerve afferents,
dmnX = dorsal motor nucleus of the vagus, io = inferior olivary
nucleus, nTS = nucleus tractus solitarii, nVII = facial nucleus,

nXII = hypoglossal nucleus, p = pyramidal tract, RS = restiform body, spV = spinal trigeminal tract, TS = solitary tract, Vagal AFF = vagal afferents, VMN = medial vestibular nucleus, IV = fourth ventricle, X = vagal efferents, XII = hypoglossal nerve.

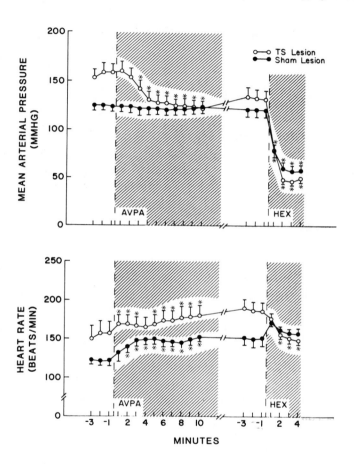

Figure 2: Comparison of the time courses of mean arterial pressure (top) and heart rate (bottom) in TS lesioned and sham lesioned dogs before and after administration of a vasopressin antagonist (AVPA) and subsequent ganglionic blockade with hexamethonium chlorid (HEX). Values are mean ± SEM. * = p< 0.05 compared to pre-drug control values.

about the localization of certain pathways in the brain facilitated the study of an animal model of neurogenic hypertension. Moreover, it appears that bilateral destruction of the solitary tracts at 5 to 7 mm anterior to the obex not only produces central interruption of the baroreflex pathway, but also spares a substantial portion of the vagal efferent fibers. One indication of a functional vagal parasympathetic outflow was the lack of significant tachycardia in the TS lesioned animals, in contrast to the tachycardia we (7) reported previously in dogs with NTS lesions which were shown histologically to impinge upon the dorsal motor nucleus of the vagus and vagal efferent fibers. Because our earlier investigation of the vagal efferent pathway in the dog medulla (19) revealed that these fibers exit the brainstem caudal to 4 mm anterior to the obex, destruction of the TS rostral to 5 mm anterior to obex can be expected to leave the vagal outflow intact. Histological examination of the TS lesions in the present study would suggest that there was minimal damage to the vagal efferent pathway.

The present pharmacological evaluation of the acute hypertension following bilateral interruption of the carotid sinus and vagal afferents in the rostral solitary tracts has revealed its primary dependence upon increased circulating vasopressin. Since the sham lesioned animals underwent the same exposure of the brain stem, evoked potential recording, and duration of anesthesia without evidence of increased vasopressin, it seems clear that the contribution of vasopressin to the hypertension in the TS lesioned dogs resulted from interruption of the baroreflex pathway. On the other hand, the similar decrease in arterial pressure following ganglionic blockade in both TS lesioned and sham lesioned animals leaves in doubt the importance of heightened activity of the sympathetic nervous system as a contributing factor to TS lesion hypertension.

Our findings agree with the observations of Sved et al. (20), who found an increase of more than twentyfold in plasma levels of vasopressin in conscious rats one hour after bilateral NTS lesions, but no rise in plasma AVP in sham lesioned animals. However, when sympathetic outflow was intact, administration of an AVP antagonist only partially reversed the hypertension in the NTS lesioned rats. Whether the difference in the contribution of AVP to the hypertension in TS lesioned dogs versus NTS lesioned rats is related to the more extensive NTS lesions in the rats, or the species difference, or both, is not clear.

In conclusion, neuroanatomical localization of the baroreceptor afferents in the dog's medulla has allowed us to produce discrete lesions of the rostral solitary tracts which interrupt the baroreflex afferent pathway. The rapid development of hypertension which is dependent upon increased circulating vasopressin following TS lesions in the dog suggests that removal of an inhibitory mechanism mediated by the sinovagal afferents may account for the release of vasopressin following TS lesions.

REFERENCES

1. Palkovits, M., Zaborsky, L. In: Hypertension and Brain Mechanism (Eds. W. DeJong, A.P. Provoost, A.P. Shapiro), Progress in Brain Research, Vol. 47, pp. 9-34,1977.
2. Doba, N., Reis, D.J. Circulation Res. 23:584-593,1973.
3. Buchholz, R.A., Nathan, M.A. Circulation Res. 54:227-238,1984.
4. Nathan, M.A., Reis, D.J. Circulation Res. 40:72-80,1977.
5. Carey, M.R., Dacey, R.G., Jane, J.A., Winn, R., Ayers, C.R., Tyson, G.W. Hypertension 1:246-254,1979.
6. Laubie, M., Schmitt, H. Am. J. Physiol. 236:H736-743,1979.
7. Ferrario, C.M., Barnes, K.L., Bohonek, S. Hypertension 3 (Suppl II): II-112-II-118,1981.
8. Kalia, M.P. In: Central Nervous System Mechanisms in Hypertension (Eds. J.P. Buckley, C.M. Ferrario), New York: Raven Press, pp. 9-24, 1981.
9. Kalia, M.P., Mesulam, M.M. J. comp. Neurol. 193:435-465,1980.
10. Ciriello, J. Neurosci. Lett. 36:37-42,1983.
11. Wallach, J.H., Loewy, A.D. Brain Res. 188:247-251, 1980.
12. Chernicky, C.L., Barnes, K.L., Ferrario, C.M., Conomy, J.P. Brain Res. Bull. 13:401-411, 1983.
13. Chernicky, C.L., Barnes, K.L., Ferrario, C.M. Neurosci. Abstr. 9: 1158,1983.
14. Mesulam, M.M. J. Histochem. Cytochem. 26:106-117,1978.
15. Barnes, K.L., Averill, D.B., Ferrario, C.M. J. Hypertension 2: 33-36,1984.
16. Ferrario, C.M., Barnes, K.L., Szilagyi, J.E., Brosnihan, K.B. Hypertension 1:235-245, 1979.
17. Winer, B.J. Statistical Principles in Experimental Design. New York: McGraw-Hill, pp. 201-204, 1971.

18. Dunnett, C.W. J. Am. Statistical Assoc. 50:1096-1121,1951.
19. Chernicky, C.L., Barnes, K.L., Ferrario, C.M., Conomy, J.P. Brain Res. Bull. 10:345-351,1983.
20. Sved, A.F., Imaizumi, T., Talman, W.T., Reis, D.J. Hypertension 7: 262-267,1985.

PART TWO

NEUROCHEMICAL BASIS
FOR CARDIOVASCULAR CONTROL

4

MORPHOLOGICAL AND BIOCHEMICAL STUDIES ON NEUROPEPTIDE Y (NPY) AND
ADRENERGIC MECHANISMS AND THEIR INTERACTIONS IN CENTRAL CARDIOVASCULAR
REGULATION

FUXE, K.(1), AGNATI, L.F.(3), HÄRFSTRAND, A.(1), KALIA, M.(4), SVENSSON
T.H.(2), NEUMEYER, A.(1), ZOLI, M.(3), LANG, R.(5), GANTEN, D.(5),
TERENIUS, L.(6), GOLDSTEIN, M.(7)

(1) Dept. of Histology and (2) Dept. of Pharmacology, Karolinska
Institutet, Stockholm, Sweden. (3) Dept. of Human Physiology, Univ. of
Modena, Modena, Italy. (4) Dept. of Pharmacology, Thomas Jefferson Univ.
Medical Center, Philadelphia, USA. (5) Dept. of Pharmacology, Univ. of
Heidelberg, Heidelberg, FRG. (6) Dept. of Pharmacology, Biochemical Center,
Univ. of Uppsala, Uppsala, Sweden. (7) Dept. of Psychiatry, New York Univ.
Medical Center, New York, USA

ABSTRACT

The adrenaline (A) neuron systems of the medulla oblongata especially
those innervating the nuc. tractus solitarius (nTS) and the nuc. dorsalis
motorius nervi vagi have been postulated by our groups to represent
important vasodepressor neuron systems. Recently morphometrical studies
have been performed characterizing the A cell groups C1 to C3 as well as
the dorsal strip A neurons of the nTS. The C1 group mainly consists of a
cluster in the rostral ventrolateral reticular formation together with
medially located scattered cells. Group C2 consists of a cluster mainly
located within the rostral part of the nuc. dorsalis motorius nervi vagi,
while group C3 mainly consists of scattered cells located in the
dorsomedial reticular formation of the rostral medulla oblongata. All of
these A cell groups together with the dorsal strip A neurons appear to
contain neuropeptide Y (NPY) immunoreactivity. Also many NPY immunoreactive
nerve terminals codistribute with A nerve terminals within various
subnuclei of the nTS. In the aged brain there is a reduction in the number
of PNMT immunoreactive nerve cell bodies and nerve terminal networks,
especially in A cell group C2. This change is parallelled by an overall
reduction of NPY immunoreactivity within these cell groups, which is even

more marked than that of PNMT immunoreactivity. Hypertension development during aging can therefore be related to a reduced synthesis and release of A and NPY in the A/NPY costoring neuron systems, particularly those belonging to the C2 group. Autoradiographical studies give evidence for the existence of a high density α-2 adrenergic binding sites within the nTS as well as a substantial number of high affinity 125 I-NPY binding sites in this region. Intracisternal injections of A, clonidine and NPY lead to hypotension, bradycardia and bradypnea. Only the effects of NPY are resistant to α-2 adrenergic receptor blockade. NPY and α-2 adrenergic receptors probably interact with one another in nTS in view of the ability of NPY to produce in vitro in membrane preparations of the medulla oblongata a reduced affinity and an increased number of binding sites for 3H-paramino-clonidine, a radioligand for high affinity α-2 adrenergic receptors. NPY also modulates adrenaline stores and/or adrenaline utilization in the nTS and VLMO region. These results suggest that one role of the adrenergic comodulator NPY may be to change the sensitivity of the α-2 adrenergic receptors via a receptor-receptor interaction, and another role may be to modulate release and/or synthesis of adrenaline. In the spontaneously hypertensive rat A release is reduced in the nTS region as well as the ability of NPY to modulate α-2 adrenergic binding sites. Also the ability of NPY to reduce arterial blood pressure is diminished in the lower dose range (25 pmol). Alterations in the NPY-/A costoring synapses of the medulla oblongata may therefore in part underlie the development of spontaneous hypertension.

INTRODUCTION

In a series of physiopharmacological studies evidence has been obtained that NPY and adrenaline (A) given intracisternally (i.c.) into the α-chloralose anaesthetized rat and intraventricularly (i.v.t.) into the awake, freely moving unrestrained rat produce marked vasodepressor and bradycardic actions (1-6). In the present paper we will summarize the evidence for the existence of interactions between NPY and adrenergic mechanisms in central cardiovascular areas using both morphological and biochemical analysis. Both presynaptic and postsynaptic interactions between NPY and adrenergic mechanisms will be considered focusing our interest on the NPY/A costoring neuronal systems. Finally possible alterations in the NPY-adrenergic mechanisms have been analyzed in relation to development of spontaneous hypertenson in the rat.

DISTRIBUTION OF NPY IMMUNOREACTIVE NEURONS OF THE MEDULLA OBLONGATA AND THE SPINAL CORD IN RELATION TO THE CATECHOLAMINERGIC NEURONS AND CYTOARCHITECTURE

Nerve Cell Bodies

NPY-like immunoreactivety has previously been demonstrated in the majority of the A nerve cell bodies of groups C1, C2 and C3 (7,8). In the present paper we have analyzed the distribution of NPY immunoreactive nerve cell bodies and nerve terminals in relation to the cytoarchitecture of the medulla oblongata, especially in regions rich in catecholamine cell bodies and nerve terminals, using the indirect immunoperoxidase procedure in combination with antibodies against NPY raised in rabbits (8). The NPY immunoreactivity of the cell bodies is enhanced by the use of colchicine treatment (120ug, i.c., 24 to 48 hours before killing). As seen in Fig. 1 to 4 NPY immunoreactive nerve cell bodies were observed in the dorsal motor nucleus of the vagus (dmnX), rostral pole (corresponding to group C2), within the dorsomedial reticular formation of the rostral part of the medulla oblongata (corresponding to group C3) and within the rostral part of the ventromedial reticular formation of the medulla oblongata located ventral to both the ventral gigantocellular reticular nucleus and to the paragigantocellular reticular nucleus (rostral level) (corresponding to group C1). At more caudal levels the NPY cell bodies were predominantly obvserved in the reticular formation lying laterally of the paragigantocellular reticular nucleus. These results are in agreement with the view that NPY-like immunoreactivity is costored with phenyl-etanolamine-N-methyltransferase (PNMT) immunoreactivity in A cell groups C1, C2 and C3. In the nucleus tractus solitarius (nTS) small NPY immunoreactive nerve cell bodies were observed within the dorsal strip (ds), in the interstitial nucleus, in the ventral and ventrolateral subnuclei and in the intermediate nucleus. Using the occlusion method (9) to quantitate coexistence, a majority of the NPY immunoreactive cells in the ds appear to contain PNMT-like immunoreactivity.

Large NPY immunoreactive nerve cell bodies are observed within the dorsal parasolitary region and within the medial subnucleus of the nTS.

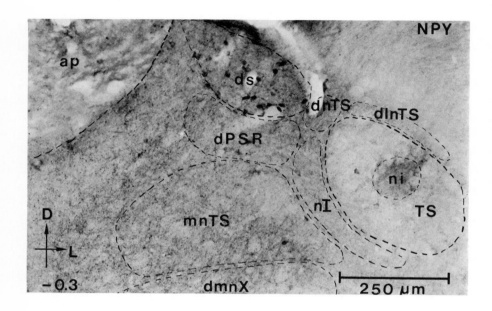

Fig. 1 NPY immunoreactivity is shown in relation to the subnuclei of the nTS of the rat medulla oblongata. The primary antiserum was diluted 1:1000. A biotinylated anti-rabbit antibody was used (1:200) followed by incubation with avidine peroxidase (1:100) (Vectastain). α-chloronaphtol was used as a substrate for the peroxidase. NPY immuno reactive cell bodies are mainly located within the dorsal strip. Abbreviations used ds = dorsal strip; dPSR = dorsal para-solitary nucleus; dnTS = dorsal subnucleus; dlnTS = dorsolateral subnucleus; mnTS = medial subnucleus; nI = intermediate nucleus; ni = interstitial subnucleus; dmnX = dorsal motor nucleus of the vagus; ap = area postrema.
Level: Obex -0.3 mm; D = dorsal; L = lateral

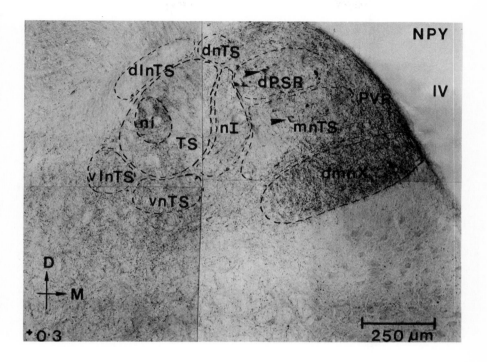

Fig. 2 NPY immunoreactivity is shown in relation to the subnuclei of the
nTS of the rat medulla oblongata. The immunoperoxidase procedure
was used as described in Fig. 1 but the diaminobenzidine (DAB) was
used as a substrate for the peroxidase. A high density of NPY
immunoreactive dots is observed various of the subnuclei. The dots
presumably represent NPY containing nerve terminals. NPY immunoreactive nerve cell bodies (arrow) are observed in the dPSR and
mnTS. Abbreviations used PVR = periventricular region. IV = 4th
ventricle. For other abbreviations see Fig. 1.
Level = obex + 0.3 mm

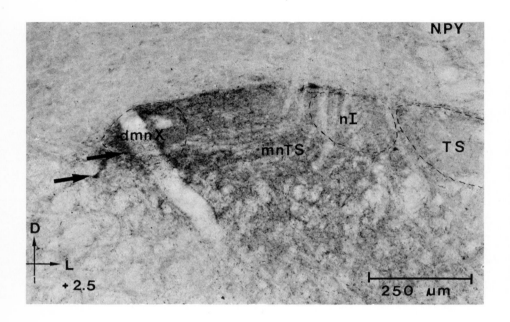

Fig. 3 NPY immunoreactivity is shown in the rostral part of the nTS by
means of the immunoperoxidase method. The same procedure as in
Fig. 1 is used. DAB was used as a substrate for the peroxidase.
Fine NPY immunoreactive puncta are observed with high densities
especially in dmnX and mnTS. This immunoreactivity is probably
located in NPY nerve terminals. NPY immunoreactive nerve cell
bodies (arrows) are observed in relation to the dmnX. The immunoreactivity of the NPY immunoreactive cell bodies is weak, because
of the lack of cholchicine treatment. For abbreviations see text
to Fig. 1.
Level = obex + 2.5 mm

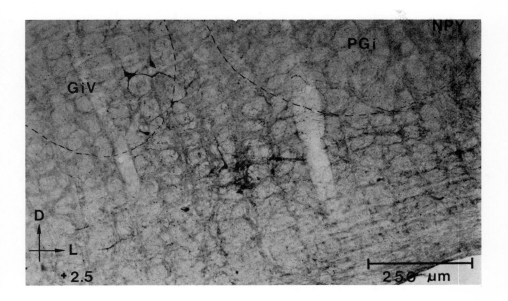

Fig. 4 NPY like immunoreactivity is shown in nerve cell bodies and
terminals in the rostral ventrolateral part of the medulla
oblongata (CI area). The immunoperoxidase procedure was used
(see Fig. 1), and DAB was employed as substrate for the
peroxidase. Abbreviations used:
GiV = gigantocellular reticular nucleus, ventral part.
PGi = paragigantocellular reticular nucleus.
Level = obex + 2.5 mm

Studies with the occlusion method indicate that some large NPY immunoreactive nerve cell bodies may also contain tyrosinehydroxylase (TH) immunoreactivity and may represent noradrenergic neurons in the A2 cell group.

Nerve Terminals

Large numbers of NPY immunoreactive nerve terminal networks were found in the dmnX and within the various parts of the nTS especially within the interstitial subnucleus of the nTS and within the rostral and caudal parts of the medial subnucleus as well as within the periventricular region and the regions surrounding the area postrema. A high degree of costorage of NPY and PNMT-like immunoreactivity has been demonstrated in nerve terminals of the medial subnucleus of nTS and in the dmnX (2).

Substantial numbers of NPY immunoreactive nerve terminals were also found within the C1 area as well as in the reticular formation lying immediately ventral to the medial subnucleus of the nTS at a rostral level. A few NPY immunoreactive nerve terminals also exist within the nuc. ambiguus, which give rise to the vagal cardiac fibres. A few NPY immunoreactive nerve cell bodies are also seen to be located immediately outside this nucleus.

Rich NPY immunoreactive nerve fibres and nerve terminals form a network surrounding the sympathetic lateral column of the spinal cord, which is also rich in TH and PNMT immunoreactive nerve terminal networks (fig. 5).

These results open up the possibilty that NPY can influence a number of cardiovascular regions of the medulla oblongata such as the nTS and C1 region. It may also directly influence the activity in the preganglionic sympathetic neurons of the sympathetic lateral column. In all these regions NPY-like immunoreactivity appears to be present both in A nerve cell bodies and terminals, and also in neuronal systems, which are structurely independent of the adrenergic systems. The observation that in fact NPY immunoreactive cell bodies exist in the ds region, especially in its dorsomedial part, where the baroreceptor afferents are known to terminate (10), is of substantial interest. It should also be considered that NPY immunoreactivity exists in the majority in the A cell bodies of group A1 (7,8) and also in some NA nerve cell bodies of group A2 (at the level immediately caudal to the obex). Thus, i.c. or i.v.t. injected NPY probably

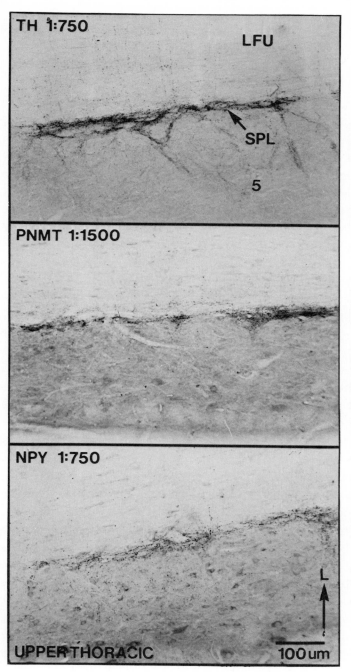

Fig. 5

TH, PNMT, and NPY-like immuno-reactivity are shown in consecutive sections of the upper part of the thoracic spinal cord of the rat. The immuno-peroxidase procedure was used (see Fig. 1) on 30 um horizontal vibratom sections. DAB was used as a substrate for the peroxidase. The dilutions are indicated. SPL = sympathetic lateral column; 5 = Rexed's lamina 5; LFU = lateral funiculus of the spinal cord; L = lateral influence a large number of NPY mechanisms in the medulla oblongata and in the spinal cord (the sympathetic lateral column).

NEUROANATOMY OF THE CATECHOLAMINE CELL GROUPS OF THE MEDULLA OBLONGATA

It was early suggested that the noradrenaline (NA) cell groups A1 and A2 of the medulla give rise to both descending connections of the spinal cord and ascending connections to the hypothalamus and the preoptic region (11-13). Retrograde tracing techniques have, however, so far only been able to demonstrate ascending connections of groups A1 and A2 into the hypothalamus and the preoptic region (14). Therefore, Kalia and colleagues (15) have used a new highly sensitive retrograde tracing technique (16) to establish the possible existence of the descending projection of the NA cell groups A1 and A2. Furthermore, in this analysis also the projection of the A cell group C1 is analyzed with regard to the projections to the hypothalamus and the spinal cord. This new procedure involves the injection of a horseradish peroxidase (HRP) - choleratoxin conjugate to facilitate their internalization into the nerve terminals of the injected region as well as a HRP - wheat germ agglutinin conjugate (WGA) to improve anterograde tracing and to protect HRP from breakdown (15,16). HRP retrogradely transported to the nerve cell bodies of the medulla oblongata were visualized after incubating the section with tetramethylbenzidine (TMB) as chromagen to form a blue product. In this way the HRP positive cells are shown to exhibit a bluish granular material in their cytoplasm. After this procedure the same section was taken to TH immunocytochemistry using the indirect immunoperoxidase procedure and diaminobenzidine (DAB) as a substrate for peroxidase. Using this procedure the double labelled cells, having TH and HRP immunoreactivity, showed a diffusely stained cytoplasm in brown (both cell bodies and dendrites) as well as a blue granular material. Using this procedure it could be demonstrated for the first time that a large part of the NA cell bodies A1 (see fig. 5) and A2 are retrogradely labelled following spinal cord injections of the tracer, giving evidence that many NA cell bodies of these two groups give rise to descending projections to the spinal cord (fig. 6) (15). In agreement with previous work injections of the tracers into the hypothalamus also produced a labelling of a large number of nerve cells in the NA cell groups A1 and A2, opening up the possibility that these neurons may give rise to both

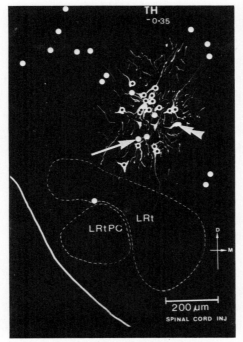

Fig. 6 Line drawing showing the tyrosine hydroxylase (TH) immunoreactive
nerve cell bodies of group A1 (noradrenaline) in the ventrolateral
part of the caudal medulla oblongata (white cells). Nerve cell
bodies, retrogradely labelled with horseradish peroxidase (HRP)
after spinal cord injection, are revealed by the use of tetramethylbenzidine as the chromagen (white dots). In the A1 cell group 3
types of nerve cell bodies are shown. The most frequent one is
represented by TH immunoreactive nerve cells, also containing
retrogradely transported HRP (white cell with dark dot). The other
2 cell types demonstrated are nerve cells, which only are TH
immunoreactive (double arrow head) and those which only are
retrogradely labelled following spinal cord injections of the
retrograde tracer (arrow).
Abbreviations used: LRt = lateral reticular nucleus; LRtPC =
parvocellular port of the reticular nucleus; D = dorsal; M =
medial. Level = obex - 0.35 mm. Solid white line to the lower left
in the fig. represents the ventrolateral border of the medulla
oblongata.

ascending and descending projections (17). It was similarly found that many cell bodies of group C1 give rise to both descending projections as well as to ascending projections to the hypothalamus. Again it is not yet known, if one and the same A nerve cell body gives rise to both descending and ascending projections into the spinal cord and also hypothalamus respectively. In previous observations (18) it seems likely that in the spinal cord both NA cell groups A1 and A2 and group C1 participate predominantly in the innervation of sympathetic lateral column (19). Thus, it seems possible that the medullary reticular NA neurons may be capable of simultaneously influencing cardiovascular centra both within the sympathetic lateral column of the spinal cord and within the hypothalamus.

MORPHOMETRICAL STUDIES ON AGING INDUCED CHANGES IN PNMT AND NPY IMMUNOREACTIVE NEURONS OF THE MEDULLA OBLONGATA

It has previously been found that peptidergic neuronal systems of the brain with few exceptions are particularly affected by the aging processes (20,21). It was therefore of interest to analyze the PNMT/NPY coexistence in the medulla oblongata of the A cell groups C1, C2 and C3 in the 24 month old animal and compare the coexistence obtained with that found in adult control rats (3 months old) from the same regions. Also it was of interest to analyze the PNMT immunoreactive neurons with regard to total immunoreactive area, dendrite area and antigen contents especially since hypertension develops in elderly and since the adrenergic neurons probably subserve a vasodepressor function (22). It was found that group C3 within the dorsomedial reticular formation of the medulla oblongata was largely unaffected by age, while the A cell group C2 located in the dmnX showed a clear cut reduction in the number of PNMT immunoreactive cells (fig. 7) associated with a parallel reduction in total immunoreactive area (FAt) and antigen contents (Ag) (23) (fig. 8). Finally, the A cell group C1 showed no reduction in number of cells but a reduction in total immunoreactive area (FAt), neuropil area (FAn) and antigen contents (Ag). (See fig. 8) The reduction in cell number in group C2 is illustrated in fig. 7. Taken together these studies on PNMT immunoreactive cell groups indicate a special vulnerability of adrenaline containing group C2 to the aging process. In addition, the C2 cell group has recently been shown by Kalia et al. (15) to give rise to efferent vagal projections to the antrum of the

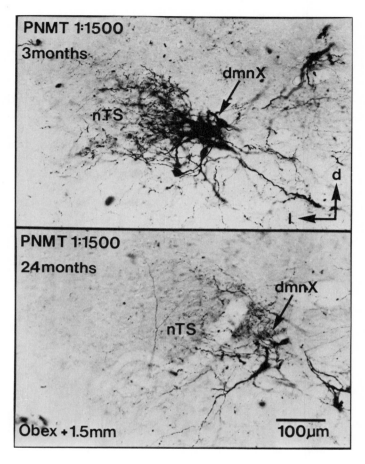

Fig. 7 PNMT-like immunoreactivity (IR) is shown in nerve cell bodies and
dendrites in the rostral pole of the dorsal motor nucleus of the
vagus (dmnX) (group C2) in the 3 and 24 month old male rat at obex
level + 1.5 mm. Transverse vibratome section (20 um). Indirect
immunoperoxidase procedure involving DAB as a substrate for the
peroxidase is used. A marked reduction of PNMT like IR is observed
in the old rat.
d = dorsal; l = lateral

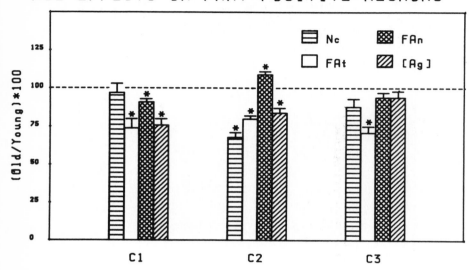

Fig. 8 The effects of aging on PNMT-positive neurons of C1, C2 and C3
cell body groups (adrenaline) measured as number of cell bodies
(Nc), total field area (FAt), field area of the neuropil (FAn)
and antigen content (Ag) using computer assisted image analysis.
Data are expressed as percentages (Old/Young) + s.e.m. (n = 3-4).
* means p<0.01. Student's t-test.

stomach.

Different results were observed when analysing the NPY immunoreactive nerve cells of the C1, C2 and C3 area of the rostral medulla oblongata. Thus, as seen in fig. 9 the 24 month old animal showed a marked reduction in the number of NPY immunoreactive cells in the C1 and C2 area, while those in the C3 area showed no changes in number in the aged rat compared with the control rat. These results indicate that in aged rats the adrenaline cell group C1 may show a preferential impairment in the mechanisms subserving the levels of the cotransmitter NPY, while the A cell group C2 undergoes a marked reduction in both NPY and PNMT immunoreactivity, suggesting a failure of both transmission lines in this system upon aging. In view of the possible vasodepressor role of groups C1 and C2, and since also NPY after central administration shows vasodepressor actions (1,5), these studies indicate that age induced changes in blood pressure regulation may not only be related to changes in peripheral mechanisms. Thus, alterations of central vasodepressor mechanisms e.g. caused by failing transmission in the A and especially in the NPY transmission lines of A group C1 and C2, should be considered. Of special interest will also be to analyze possible quantitative changes in the ds A/NPY neurons in view of the localization of these putative interneurons in the ds, which is richly innervated of baroreceptor afferents.

STUDIES ON THE SITE OF ACTION OF THE VASODEPRESSOR EFFECTS OF NPY, A AND CLONIDINE: A RECEPTOR AUTORADIOGRAPHICAL ANALYSIS

In the receptor autoradiographical experiments the radioligand 3H-para-aminoclonidine (3H-PAC, specific activity 40 Ci/mmol, NEN) was used to label the α_2 agonist binding sites and ^{125}I-NPY (Amersham, G.B., specific activity 2000 Ci/mmol) to study the distribution of NPY recognition sites. In the experiments on 3H-PAC the biochemical procedure of Rouot and Snyder (24) was followed and in the experiments on ^{125}I-NPY binding, the biochemical procedure of Goldstein et al. (25) was followed. In the experiments on iodinated NPY the Tris hydrochloride buffer was used containing 5 mM magnesium chloride and calcium chloride as well as 0.1 mM bacitracin and 0.5 % bovine serum albumine. Using a concentration of 0.5 nM, a concentration which is close to the K_D value of the NPY binding sites, a high degree of labelling was demonstrated in the nTS, in the parasolitary area and in the inferior olive (see fig. 10). With this low concentration of iodinated NPY areas are demonstrated, which contain a high density of iodinated NPY binding sites. Studies on the distribution of 3H-PAC also demonstrate a high labelling of the nTS region and the parasolitary region (see fig. 11). As shown a substantial overlap exists between the iodinated NPY labelled area and the 3H-PAC labelled area within the nTS, which includes also the dmnX. Taken together these receptor autoradiographical studies indicate that both NPY and α_2 agonists may exert part of their vasodepressor effects in the nTS region. The sympathetic lateral column of the spinal cord should, however, also be considered as a site of action.

STUDIES ON INTERACTIONS BETWEEN α_2 AND NPY RECEPTORS: BIOCHEMICAL AND RECEPTORAUTORADIOGRAPHICAL STUDIES

This work has demonstrated that NPY (10 nM) in vitro can increase the number of 3H-PAC sites in membranes preparations from the medulla oblongata as well as the K_D value of these binding sites (26). In subsequent work it could also be demonstrated that a similar modulation was also induced by NPY in vitro in the α_2 adrenergic antagonist binding sites as revealed in studies using the radioligands 3H-rauwolscine and 3H-idazoxan (RX781094) (27). In the present study it can also be demonstrated (see figs. 12, 13) that NPY can modulate 3H-PAC binding sites, when using mock CSF as incubation medium. Thus, an increased density is observed in the presence of 10 nM of NPY (fig. 12) and in the presence of 1 nM of NPY a selective increase of the K_D value is noted (fig. 13). In receptorautoradiographical studies evidence was also obtained that NPY reduced the affinity of 3H-PAC binding sites in the nTS area and the parasolitary region of the medulla oblongata using mock CSF as incubation medium. In these studies the concentration used of NPY was 10 nM (2). These studies taken together indicate that NPY receptors by receptor-receptor interactions may modulate the binding characteristics of α_2 agonist and antagonist binding sites in inter alia cardiovascular areas of the medulla oblongata, especially the nTS.

As seen in fig 14. it has also been possible by means of receptorautoradiography to show that clonidine (10 nM) can modulate the binding characteristics of ^{125}I-NPY binding in sections of the medulla

EFFECTS OF AGING ON CELL NUMBER
ROSTRAL MEDULLA OBLONGATA

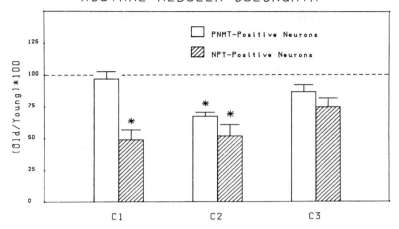

Fig. 9 Age related changes in the number of NPY- and PNMT-positive cell
bodies in the rostral medulla oblongata (C1, C2 and C3 groups,
respectively). Means $\pm$ s.e.m. of the percentages (young/old) are
given (n = 3-4). Student's t-test.
* = p<0.01

Fig. 10 Autoradiographical localisation of ^{125}I-NPY binding sites in
coronal section (14 um) of the rat medulla oblongata using a
tritium sensitive film (^{3}H-Ultrofilm LKB, Sweden). Concentration
used was 0.5 nM. The ^{125}I-NPY (Amersham G.B.) had a specific
activity above 2000 Ci/mmol. The experiments were performed during
optimal equilibrium conditions using a 50 mM Tris HCl buffer (pH
7.6) containing 5 mM $MgCl_2$ and $CaCl_2$ as well as 0.1 mM bacitracin.
Incubation time was 90 minutes at room temperature. Unspecific
binding was defined as binding in the presence of 1 uM NPY. In the
lower panel the neuroanatomical landmarks are shown based on a
coronal section from the Atlas of Paxinos and Watson (34). The
binding measured by quantitative receptor autoradiography was
studied in the areas out-lined by solid lines. As seen the nTS
area includes the medial nucleus of the tractus solitarius and the

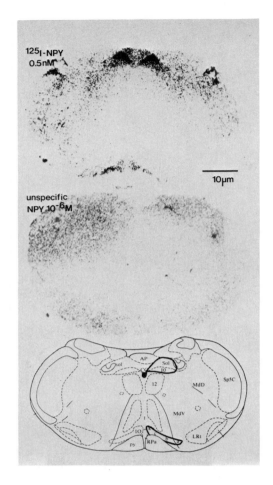

dmnX. The parasolitary area includes mainly the lateral nucleus of
the tractus solitarius (SolL) and the parasolitary nucleus (PSol).
The area in the inferior olive includes the medial nuclei as well
as the dorsal nucleus. SOL = tractus solitarius; IO = inferior
olive; RPa = nucleus raphe pallidus; py = pyramidal tract; XII =
nucleus hypoglossus; MdV = ventral reticular nucleus of the medulla
oblongata; MdD = dorsal reticular nucleus of the medulla oblongata;
LRt = lateral reticular nucleus; AP = area postrema; Sp5 = spinal
nucleus of the trigeminal nerve; PSol = parasolitary nucleus;
PCRt = parvocellular reticular nucleus of the medulla oblongata

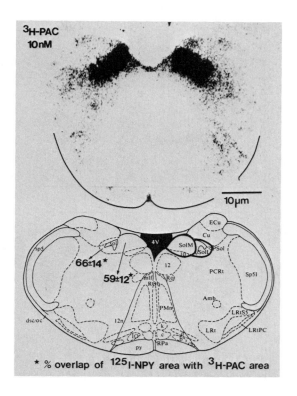

Fig. 11 Autoradiograpical localisation of ^{3}H-para-aminoclonidine (PAC)
binding sites (10 nM, specific activity 40 Ci/mmol) in a coronal
section (14 um) of the rat medulla oblongata using tritium
sensitive film (^{3}H-Ultrofilm LKB, Sweden). In the lower part the
neuroanatomical landmarks are shown. By means of a computer
assisted analysis the overlap areas of ^{3}H-PAC area and ^{125}I-NPY
area were analyzed. The % overlap is shown. The overlap regions
are indicated by solid lines in the lower panel of the fig. The nTS
area includes the medial subnucleus (solM) and the dmnX (10) and
the parasolitary area includes the lateral nucleus of the tractus
solitarius (SolL) and the parasolitary nucleus (PSol).

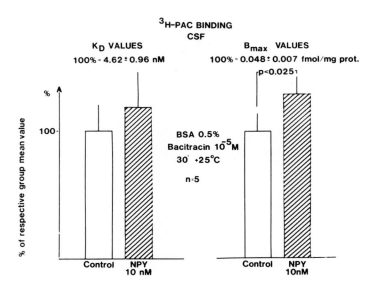

Fig. 12 Effects of NPY (10nM) in vitro on the binding characteristics of
^{3}H-PAC in membrane preparations of the male rat medulla oblongata
(obex - 1mm to obex + 3 mm) using mock CSF (CSF) as incubation
medium. The experiments were performed under equilibrium conditions
with an incubation time of 30 min. at room temperature. Bovine
serum albumin (BSA) (0.5 % w/v) and bacitracin (10^{-5}M) were also
present and optimized the procedure. Means ± s.e.m. are shown and
the values are expressed in % of respective control group mean
value (CSF alone). (n = 5 experiments). Student's paired t-test
was used.

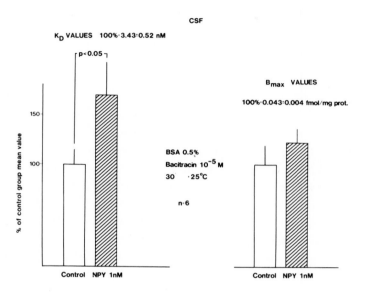

Fig. 13 Effects of NPY (1 nM) in vitro on the binding characteristics of ^{3}H-PAC in membrane preparations from the rat medulla oblongata (obex - 1mm to + 3 mm). For further details, see text to Fig. 8. Means ± s.e.m. are shown and the values are expressed in % of respective control group mean value (CSF alone). n = 6 experiments. Student's paired t-test was used in the statistical analysis.

oblongata. Thus, clonidine induces a reduced binding of iodinated NPY (0.5 nM) in the nTS and in the parasolitary region and also in the inferior olive. These results indicate that α_2 adrenergic receptor can also modulate NPY binding. Thus, a bidirectional receptor-receptor interaction may exist between these 2 types of receptors in membranes of the region of the nTS and parasolitary area.

Also the pharmacological studies on the effects of NPY and A on arterial blood pressure upon central administration indicate the existence of interactions between NPY and α_2 adrenergic receptor mechanisms (5). These cardiovascular studies indicate the possibility that these 2 receptor mechanisms may interact at the coupling device level probably at the level of the inhibitory GTP binding Niα protein, which induces inhibition of adenylate cyclase activity. Such an interaction at this level may secondarily lead to changes in the binding characteristics of the proteins carrying the NPY and α_2 agonist binding sites, respectively. However, the fact that no additive effects were observed between A and NPY, when they were administered together even when using threshold doses also indicate an existence of interactions at the level of the protein containing the recognition sites. Thus, when e.g. the NPY receptor population is activated it may form a cluster with the α_2 receptors, which may be less coupled leading to the existence of a rapidly developed off-switch mechanism for the activity at the α_2 agonist sites (see fig. 15).

PRESYNAPTIC EFFECTS OF INTRAVENTRICULARLY ADMINISTERED NPY ON REGIONAL A LEVELS AND UTILIZATION

As seen in fig. 16 NPY in a high dose of 1.25 nmol given i.v.t. in the awake unrestrained freely moving male rat produced a depletion of the A stores in the caudal part of the nTS region (caudal DCMO) and a trend for an enhancement of the A depletion, induced by the PNMT inhibitor LY134046 (40mg/kg, i.p. 4 hours before killing). Instead, in the thoracic part of the spinal cord and in the hypothalamus NPY produced a significant reduction in the depletion of the A stores by LY134046. These results indicate that NPY directly or indirectly can enhance A utilization and reduce A stores in the caudal part of the nTS and reduce A utilization in the hypothalamus and in the thoracic part of the spinal cord (3). Interpretation of these results is difficult, since they may well be

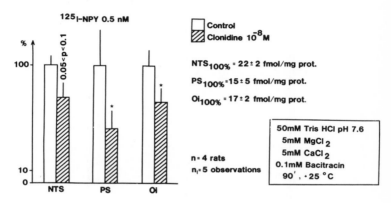

EFFECTS OF CLONIDINE (10 nM) 125 I-NPY BINDING IN SECTIONS OF THE MEDULLA OBLONGATA USING RECEPTOR AUTORADIOGRAPHY

Fig. 14 Effects of clonidine (10 nM) on ^{125}I-NPY binding in coronal
sections (14 um) of the rat medulla oblongata using quantitative
receptor autoradiography. The sections were analyzed from the
level 0.5 mm caudal to the obex to a level 1 mm rostral to the
obex. An IBAS (Zeiss Kontron, FRG) was used as a densitometer.
The concentration used was 0.5 nM of iodinated NPY. For binding
procedures, see Goldstein et al. (25). Means $\pm$ s.e.m. are shown
in % of respective NPY control group mean value. n = 4 rats, 5
observations/rat. nTS = nucleus tractus solitarius area, which
includes also the dmnX; PS = parasolitary area which includes
the lateral solitary nucleus; Oi = inferior olive.
Student's paired t-test was used for the statisitical analysis.
* = p $<$ 0.05

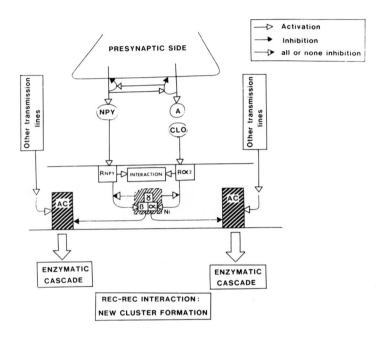

Fig. 15 A schematic illustration of interactions between NPY and α2
adrenergic receptor at the level of the pre- and postsynaptic
membrane of NPY/A synapses of the medulla oblongata of the rat.
CLO = clonidine

secondary to the blood pressure lowering action of NPY. As a matter of fact this may be one possible mechanism for the ability of NPY to reduce A release and utilization in the projection areas of predominantly A cell group C1. Thus, these adrenergic projections probaly have a vasodepressor function (28-30). This adrenergic neuronal system must therefore be turned off in order to avoid a dangerous and fatal lowering of the arterial blood pressure, so that the animal may survive. NPY receptors may thus, within the medulla oblongata and the spinal cord via direct or indirect connections at the network level turn off activity in the adrenergic neurons.

The increase of A utilization in the caudal part of the nTS and the depletion of A stores observed in this area following NPY treatment may possibly be related to the existence of ds adrenergic neurons, which probably represent local neurons terminating within the nTS itself. It seems e.g. possible that NPY given i.c. inhibit activity in vasopressor projection neurons from the nTS, which in turn via inhibitory recurrent collaterals control activity in the local ds adrenergic neurons, which in turn are controlled by baroreceptor afferents. These neurons probably have a vasodepressor function. Thus, when the NPY receptor activation in these regions of the nTS leads to a loss of the recurrent inhibitory feed back loop, an increase of A utilization in this region is induced. Such a mechanism may also allow for a high gain of the baroreceptor reflex (3).

It is also of substantial interest that in release experiments on synaptosomes of the medulla oblongata it has been found that NPY (1 nM) can enhance the inhibitory effects of clonidine on 3H NA release (31). Thus, there may also exist presynaptic NPY binding sites on the A and NA nerve terminals of the medulla oblongata, which may enhance the adrenergic autoreceptor function, leading to inhibition of NA and A release in the NA and A nerve terminals, respectively. Such a mechanism (31) may contribute to the reduction of A utilization observed in the hypothalamus and the spinal cord upon i.v.t. administration of NPY.

Indications also exist that clonidine can modulate presynaptic NPY mechanisms in discrete regions of the medulla oblongata. Thus, clonidine (3.75 nmol/rat) can significantly reduce NPY-like immunoreactivity 5 minutes and 1 hour following i.c. injections into the α-chloralose anaesthetized male rat within the rostral part of the nTS (rostral DCMO) (32). Instead the caudal part of nTS region shows no reduction in NPY-like

74

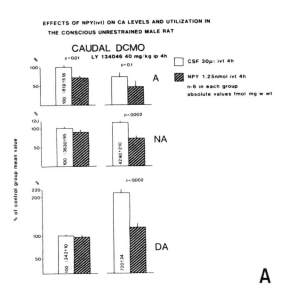

EFFECTS OF NPY(ivt) ON CA LEVELS AND UTILIZATION IN
THE CONSCIOUS UNRESTRAINED MALE RAT

A

Fig. 16 A,B Effects of NPY (1.25 nmol i.v.t.) on CA levels and utilization
in conscious, unrestrained male rats in discrete regions of
the rat central nervous system. Mock CSF was used as solvent
(CSF). Absolute values indicated in the columns are expressed
in fmol/mg w.w.. The following regions were studied within the
caudal part of the dorsomedial medulla oblongata: caudal dcmo
(A) = obex + 1 mm; rostral dcmo obex + 1 to + 3 mm; ventrolateral part of the reticular formation of the medulla
oblongata (vlmo B). The vlmo extends 4 mm in the caudal-rostral
direction from 1 mm caudal to obex to 3 mm rostral to obex. In
this area both NA cell group A1 and A cell group C1 are
located. In the studies on the effects of NPY on A utilization
the depletion of A was studied by means of the PNMT inhibitor
LY134046 (40mg/kg i.p. in 0.3 ml saline, 4 h before killing).
NPY was given immediately after the injection of LY134046.
Catecholamine levels in brain tissue were determined by means

of a high pressure liquid chromatography in combination with electrochemical detection. Means $\pm$ s.e.m. are shown in % of control group mean value (CSF i.v.t., saline i.p.). The statistical analysis was carried out by means of Mann-Whitney U-test (nonparametrical).

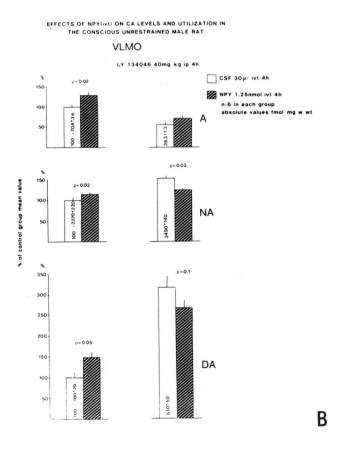

EFFECTS OF NPY(ivt) ON CA LEVELS AND UTILIZATION IN
THE CONSCIOUS UNRESTRAINED MALE RAT

VLMO

immunoreactivity at these 2 time intervals following an i.c. injection of clonidine (3.75 nmol). These presynaptic effects of clonidine may reflect reduced synthesis of NPY-like immunoreactivity or more likely an enhanced NPY release. Following chronic clonidine treatment (100 ug/kg, s.c. twice daily, 14 days) tolerance has developed with regard to the depleting action of clonidine on NPY-like immunoreactivity in the rostral part of the nTS.

CHANGES IN NPY AND ADRENERGIC MECHANISMS IN THE SH RAT COMPARED WITH THE WISTAR-KYOTO NORMOTENSIVE RAT

In previous experiments it has been shown that there exists a reduced A utilization in the area of the nTS of the SH rat compared with the normotensive Wistar-Kyoto rat (29). Also immobilization stress has been found to induce depletions of A stores in the nTS area of the SH rat (22) but not of the Wistar-Kyoto rat. These results indicate disturbances in the mechanisms contolling A synthesis and release in the adrenergic neuronal networks of the nTS (22). Recently, these studies on immobilization stress have been continued (33). In these experiments it could be demonstrated that chronic immobilization stress induced a depletion of A stores in the rostral part of the nTS, exclusively in the SH animal. It was also shown that in the spinal cord (thoracic part) chronic immobilization stress led to a marked increase of the A levels. Studies with dopamine ß-hydroxylase inhibitors indicate that this increase in A levels is caused by a reduction of A release. These results underline the view that the descending adrenergic system to the sympathetic lateral column may be a vasodepressor system which cannot be appropriately activated during chronic immobilization stress in the SH rat but instead shows an abnormal reduction of A release and utilization upon exposure to such a stress (33).

Studies on NPY-like immunoreactivity in various regions of SH rat compared with the Wistar-Kyoto normotensive rat show a trend for a reduction of NPY-like immunoreactivity in the rostral part of the nTS and a significant reduction of NPY-like immunoreactivity within the somatosensory area of the frontoparietal cortex (see fig. 17). However, the other areas analyzed, that is the caudal part of the nTS locus coeruleus and the anteromedial frontal cortex did not show any changes in the levels of NPY-like immunoreactivity. Thus regionally selective changes in the NPY immunoreactive stores appear to exist in the SH compared with the

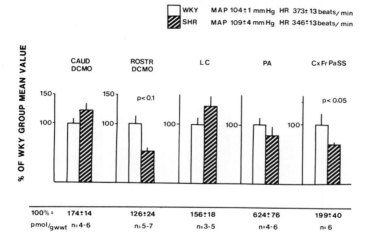

Fig. 17 NPY-like immunoreactivity is shown in various discrete brain
regions of the 4 week old SH and WKY male rat. The radioimmuno
assay technique was used for the analysis. Abbreviations used LC =
locus coeruleus; PA = peri- and paraventricular hypothalamic
region; CxFrPaSS = somatosensory area of the frontoparietal
cortex. The 100 % values are expressed in pmol/g w.w.. In separate
experiments mean arterial blood pressure (MAP), heart rate (HR)
were measured. No statistical difference in the cardiovascular
parameters was obtained. For statistical analysis Mann-Whitney
U-test (nonparametrical) was used.

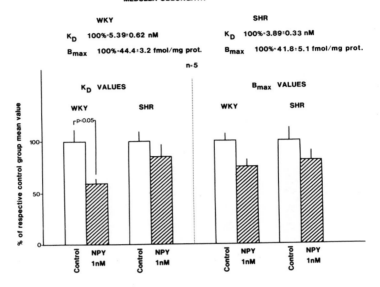

Fig. 18 The effects of NPY (1 nM) in vitro on the binding characteristics of ^{3}H-PAC in membrane preparations of the medulla oblongata of the adult male WKy rat and of the adult male SH rat are shown. Mock CSF (CSF) was used as incubation medium. Bovine serum albumin (0.5 %) and bacitracin (10 um) were also present in order to optimize conditions. Experiments were performed under optimal equilibrium conditions. Incubation time was 30 min at 25° C. Means ± s.e.m. are shown in % of respective control group mean value. (n = 5 experiments). Student's paired t-test was used. It must be pointed out that in the WKy rat NPY (1 nM) enhances the affinity while in the Sprague-Dawley rat it reduces the affinity of the α_2 adrenergic agonist binding sites, when using mock CSF as incubation medium.

Wistar-Kyoto rat. Furthermore, studies on the cardiovascular actions of NPY in the SH rat indicate a reduced potency of the i.c. administered NPY to reduce arterial blood pressure and heart rate in the SH rat compared with the Wistar-Kyoto rat (4). These results may be explained by the existence of a reduced affinity of the NPY binding sites of the SH animal. In line with these results it has been found that NPY (1 nM) in vitro can no longer modulate the paraminoclonidine binding sites in vitro in membrane preparations of the medulla oblongata of the SH animal (see fig. 18).

Taken together these results indicate several neurochemical changes in the adrenergic and NPY neurons of the SH rat compared with the Wistar-Kyoto rat. Of special interest are the marked alterations in the stress reactivity of the adrenergic neurons of the medulla oblongata and the spinal cord in response to chronic immobilization stress.

ACKNOWLEDGEMENTS

This work has been supported by a grant (04X-715) from the Swedish Medical Research Council, by a grant from Svenska Läkarsällskapet and by a grant from NIH. We are grateful to Ms Anne Edgren, Ms Rose Marie Gustafsson for excellent secretarial service.

REFERENCES

1. Fuxe, K., Agnati, L.F., Härfstrand, A., Zini, I., Tatemoto, K., Merlo Pich, E., Hökfelt, T., Mutt, V., Terenius, L. Acta physiol. scand. 118:189-192,1983.
2. Fuxe, K., Agnati, L.F., Härfstrand, A., Janson, A.M., Neumeyer, A., Andersson, K., Ruggeri, M., Zoli, M., Goldstein, M. Prog. Brain Res.:in press, 1985a.
3. Fuxe, K., Härfstrand, A., Agnati, L.F., Kalia, M., Neumeyer, A., Svensson, T.H., Von Euler, G., Terenius, L., Bernardi, P., Goldstein, M. In: International Symposium on brain Epinephrine, Sept. 29-Oct. 3, Baltimore, USA. (Eds. J.M. Stolk, D. U'Prichard & K. Fuxe), in press, 1985b.
4. Härfstrand, A., Fuxe, K., Agnati, L.F., Ganten, D., Eneroth, P., Tatemoto, K., Mutt, V. Clin. exp. Hypertens., Theory and Practice A6 (10-11):1947-1950,1984.

5. Härfstrand, A. et al. this symposium.

6. Borkowski, K.R., Finch, L. In: Central Adrenaline Neurons, Basic aspects and their role in cardiovascular functions (Eds. K. Fuxe, M. Goldstein, B. Hökfelt & T. Hökfelt), Pergamon Press, Oxford and New York, pp. 225-234,1980.

7. Hökfelt, T., Lundberg, J.M., Tatemoto, K., Mutt, V., Terenius, L., Polak, J., Bloom, S., Elde, R., Goldstein, M. Acta physiol. scand. 117:315-318,1983.

8. Everitt, B.J., Hökfelt, T., Terenius, L., Tatemoto, K., Mutt, V., Goldstein, M. Neuroscience Vol. 11, No. 2:443-462,1984.

9. Agnati, L.F., Fuxe, K., Locatelli, V., Benfenati, F., Zini, I., Panerai, A.E., El Etreby, M.F., Hökfelt, T. J. Neuroscience Methods 5: 203-214,1982.

10. Spyer, K.M., Donoghue, S., Felder, R.B., Jordan, D. Clin. exp. Hypertens.-Theory Pract. A6 (1 and 2):173-184, 1984.

11. Dahlström, A., Fuxe, K. Acta physiol. scand. Vol 64, Suppl. 247: 7-34,1965.

12. Fuxe, K. Acta physiol. scand. Vol. 64, Suppl. 247:39-85,1965.

13. Anden, N.E., Dahlström, A., Fuxe, K., Larsson, K., Olson, L., Ungerstedt, U. Acta physiol. scand. 67:313-326,1966.

14. Lindvall, O., Björklund, A. In: Chemical Neuroanatomy (Ed. P.C. Emson), Raven Press, New York. pp. 229-255,1983.

15. Kalia, M., Fuxe, K., Gelb, D., Hudson, M., Härfstrand, A., Goldstein, M. In: 10th International Symposium on G.I. Motility, 1985b.

16. Wan, X.-C., Trojanowski, J.Q., Gonatas, J.O. Brain Res. 243:215-224, 1982.

17. Olson, L., Fuxe, K. Brain Res. 43:289-295,1972.

18. Nygren, L.G., Olson, L. Brain Res. 132:85-93,1977.

19. Hökfelt, T., Fuxe, K., Goldstein, M., Johansson, O. Brain Res. 66: 235-251,1974.

20. Agnati, L.F., Fuxe, K. Biosci. Rep. 4:93-98,1984a.

21. Agnati, L.F., Fuxe, K., Benfenati, F., Toffano, G., Cimino, M., Battistini, N., Calza, L., Merlo Pich, E. Acta physiol. scand., Suppl. 532:45-61,1984b.

22. Fuxe, K., Agnati, L.F., Ganten, D., Goldstein, M., Yukimura, T., Jonsson, G., Bolme, P., Hökfelt, T., Andersson, K., Härfstrand, A. Unger, T., Rascher, W. In: Central Nervous System Mechanisms in

Hypertension (Eds. J.P. Buckley, C.M. Ferrario), Raven Press, New York, pp. 89-113, 1981.

23. Agnati, L.F., Fuxe, K., Grimaldi, R., Härfstrand, A., Zoli, M., Zini, I., Ganten, D., Bernardi, P. In: The molecular basis for the central and peripheral regulation of the vascular resistance (Eds. W. Osswald, D. Reis, P. Vanhoutte), Plenum Press, 1985.

24. Rouot, B., Synder, S.H. Life Sci. 25:769-774,1979.

25. Goldstein, M., Kusano, N., Adler, C., Meller, E. Prog. Brain Res. Elsevier Science Publ., Amsterdam, in press, 1985.

26. Agnati, L.F., Fuxe, K., Benfenati, F., Battistini, N., Härfstrand, A., Tatemoto, K., Hökfelt, T., Mutt, V. Acta physiol. scand. 118:293-295, 1983.

27. Fuxe, K., Agnati, L.F., Härfstrand, A., Martire, M., Goldstein, M., Grimaldi, R., Bernardi, P., Zini, I., Tatemoto, K., Mutt, V. Clin. exp. Hypertens. A6(10&11):1951-1956,1984.

28. Fuxe, K., Hökfelt, T., Bolme, P., Goldstein, M., Johansson, O., Jonsson, G., Lidbrink, P., Ljungdahl, A., Sachs, CH. In: Central Action of Drugs in Blood Pressure Regulation (Eds. D.S. Davies and J.S. Reis), Pitman Medical, London, pp. 8-22, 1975.

29. Fuxe, K., Ganten, D., Jonsson, G., Agnati, L.F., Andersson, K., Hökfelt, T., Bolme, P., Goldstein, M., Hallman, H., Unger, T., Rascher, W. Neurosci. Lett. 15:183-188,1979

30. Franz, D. In: International symposium on brain epinephrine, Baltimore Sept. 29 - Oct. 3. (Eds. J.M. Stolk, D. U'Prichard and K. Fuxe), 1985.

31. Martire, M., Fuxe, K., Pistritto, G., Preziosi, P., Agnati, L.F. J. neural Transm., submitted, 1985.

32. Härfstrand, A., Fuxe, K., Lang, R.E., Ganten, D. unpublished data

33. Svensson, T.H., Härfstrand, A., Fuxe, K., Ganten, D. In: International Symposium on Brain Epinephrine, Sept. 29 - Oct. 3, Baltimore, USA. (Eds. J.M. Stolk, D.U'Prichard and K. Fuxe), 1985.

34. Paxinos, G., Watson, C. The rat brain in stereotaxic coordinates. Academic Press, 1982.

5

TYROSINE'S   EFFECTS   ON   BLOOD   PRESSURE   DURING   HYPOTENSION

CONLAY, L.A., MAHER, T.J., WURTMAN, R.J.

Department of Anesthesia, Massachusetts General Hospital, and Harvard Medical School, Fruit Street, Boston, Massachusetts 02114

ABSTRACT

Hemorrhagic shock is accompanied by an intense activation of the sympathetic nervous system, mediated, at least in part, by the release of sympathoadrenal catecholamines (1). Tyrosine is the amino acid precursor of these catecholamine neurotransmitters (norepinephrine, epinephrine, and dopamine), and its administration increases both catecholamine synthesis and release when catecholamine-containing neurons fire frequently (2-4). For example, tyrosine increases catecholamine synthesis and release during cold stress (5), increases dopamine release ipsilateral to partial nigrostriatal lesions (6), and prevents reserpine-induced hyperprolactinemia (7).

During hemorrhagic shock, catecholamine-containing neurons in the periphery (1) and spinal cord (8) fire frequently, causing vasoconstriction and increases in heart rate. Likewise, tyrosine increases blood pressure (BP) by increasing catecholamine synthesis (9-12). This chapter will discuss the mechanism by which tyrosine, and neurotransmitter precursors in general, increase neurotransmitter synthesis; the mechanism of tyrosine's pressor activity during hemorrhagic shock; and the possible therapeutic effects of tyrosine and other large neutral amino acids on sympathetic outflow.

1.   BIOCHEMICAL CRITERIA FOR PRECURSOR CONTROL

1.1 General Concepts. Neurotransmitters are divided by chemical structure into three general categories: 1) amines (catecholamines, serotonin, acetylcholine, and histamine); 2) peptides (opiates, substance P, thyrotropin releasing factor, etc); and 3) nonessential amino acids and their metabolites (glycine, GABA or gamma-amino butyric acid, glutamate)

(3,12,13). The amine group of neurotransmitters (1) is unique, in that their synthesis and release are influenced by the concentration of their biochemical precursors, compounds that are consumed in the diet. In general, precursors may influence the rate of neurotransmitter synthesis when several conditions are satisfied (3,12). First, the neurotransmitter's biochemical precursor must cross the blood-brain barrier and gain entry into the neuron. Second, the neuron must be able to generate more product when exposed to more of its precursor substrate. Third, the neurotransmitter product must not feed back to inhibit its own synthesis.

The precursors for amine neurotransmitters cross the blood-brain barrier and other cell membranes by facilitated diffusion, through a specific carrier transport system (14). Moreover, the enzymes that catalyze amine neurotransmitter synthesis are not saturated with substrate at normal tissue concentrations (3,4,12). Since all the enzymatic binding sites are not occupied, the addition of more substrate generates more neurotransmitter product. Finally, feedback inhibition has not been described for serotonin or acetylcholine, and appears unimportant in rapidly-firing catecholamine-containing cells (3,12).

1.2 Tyrosine is the amino acid precursor of the catecholamines. It can be derived either directly from the diet mostly in the form of proteins, from hepatic metabolism of the amino acid phenylalanine, and/or from lysis of endogenous proteins (3). Tyrosine occupies an amino acid carrier system within the blood brain barrier shared with phenylalanine, valine, leucine, isoleucine and tryptophan (14). Since these large neutral amino acids (LNAAs) share a common carrier, it is the ratio of tyrosine to LNAA, rather than the plasma tyrosine concentration alone, that determines brain tyrosine concentration (15).

Tyrosine hydroxylase is the rate-limiting enzyme in catecholamine synthesis, and it is unsaturated with its amino acid substrate at normal tyrosine concentrations (3,4). However, the ability of tyrosine to increase catecholamine synthesis is specific for frequently-firing neurons (3,12,13). In resting catecholamine-containing cells, tyrosine hydroxylase, the rate-limiting enzyme in catecholamine synthesis, is affected by end product inhibition and is unsaturated with its cofactor, so tyrosine administration under resting conditions does not increase catecholamine synthesis (16,17). When a catecholamine-containing neuron fires frequently, the tyrosine hydroxylase enzyme is phosphorylated, resulting in kinetic changes that increase the enzyme's saturation with cofactor and diminish its susceptibility to end-product inhibition. Under conditions of frequent neuronal firing, the tyrosine concentration becomes the limiting factor in the catecholamine synthesis sequence. Therefore, tyrosine administration to frequently firing neurons generates more of the catecholamine product (3,4,16).

2. EXPERIMENTAL MODEL

Experiments used male Sprague-Dawley rats (retired breeders obtained from Charles River Laboratories), weighing approximately 500 g (9). The animals were anesthetized with alpha-chloralose (50 mg/kg i.p.) and urethane ethyl carbamate (500 mg/kg), and tracheostomies were performed. The vagus nerve and the cervical sympathetic trunk were seperated from the left carotid artery, and the vessel was cannulated with PE 50 tubing. BP was continuously recorded (except while drugs were being administered intra-arterially) using a Grass Model 70 polygraph and Statham pressure transducers. Hemorrhagic hypotension was induced by bleeding each animal until its systolic BP was reduced to half the starting value, approximately 50 mm/Hg; blood was withdrawn in 1 ml increments over a 5-min period. Pretreatments were administered as described. Tyrosine and tyramine were administered after animals had been hypotensive or normotensive for one hour.

3. TYROSINE INCREASES BLOOD PRESSURE IN HYPOTENSIVE ANIMALS

Tyrosine's administration to the hypotensive animals caused a dose-related increase in systolic blood pressure (9). This might have been predicted, since catecholamine-containing neurons fire frequently during hemorrhage: in the periphery, sympatho-adrenal neurons release catecholamines, causing vasoconstriction and heart rate increases during an hemorrhagic insult (5).

Evidence supporting catecholamine synthesis as tyrosine's mechanism of action was derived from three sources (10). First, tyrosine's ability to increase BP was blocked when catecholamine synthesis in peripheral tissues was blocked with Carbidopa. Second, tyrosine failed to increase BP in hypotensive rats pretreated with an alpha blocker, phentolamine. Third, adrenalectomy blocked tyrosine's pressor effect in hypotensive animals, and tyrosine's administration increased adrenal epinephrine concentrations (an effect not previously noted in normal animals) (8, 10). The amino acid's ability to increase catecholamine synthesis after a period of hypotension is consistent with the view that neurons become responsive to the amino acid only after a period of accelerated neuronal firing, which has activated tyrosine hydroxylase.

3.1 Tyrosine's pressor effect in hypotensive rats is not mediated by tyramine. Tyrosine may be decarboxylated in vivo to form tyramine, an indirect sympathomimetic amine (18). Though some authors have suggested that very little tyramine is actually formed when rats are given tyrosine, others have presented evidence that such conversion is quantitatively significant (19-22). The physiological significance of tyrosine's decarboxylation to tyramine is yet to be defined, though this mechanism has been proposed both for tyrosine's ability to decrease BP in hypertensive rats, and for the amino acid's antidepressant effect (21). If tyrosine were decarboxylated to tyramine in hypotensive rats, then its pressor effect could be mediated by the release of stored norepinephrine from sympathetic terminals, rather than by the synthesis of additional catecholamine molecules as previously suggested. Moreover, a tyramine-mediated release of stored norepinephrine might be expected to cause tachyphylaxis, and eventually death from catecholamine depletion.

Three sets of experiments provided evidence that different mechanisms mediated the pressor actions of these two compounds (23). First, pretreatment with reserpine, an indirect sympathomimetic amine, blocked tyramine's pressor effect, but not that of tyrosine. (Tyramine releases endogenous catecholamine stores; since reserpine depletes these stores, it blocks tyramine's effect). Second, while reserpine blocked tyramine's but not tyrosine's ability to increase BP, ganglionic blockade with hexamethonium had the opposite effect. Tyrosine failed to increase BP in animals made hypotensive with hexamethonium, though it did exert a pressor effect in animals made hypotensive by hemorrhage. Conversely, tyramine increased BP both in hexamethonium-treated animals and in those made hypotensive by hemorrhage. These findings are consistent with an indirect sympathomimetic effect for tyramine, which should not have been affected by ganglionic blockade, as well as a tyrosine-induced increase in catecholamine synthesis. Finally, tyramine was undetectable in plasma of rats receiving even a large dose of tyrosine (100 mg/kg, which increased BP by 39 +/- 7 mm Hg) (23). Therefore, the pressor effect of tyrosine in hypotensive animals was probably not the result of the amino acid's decarboxylation to tyramine.

3.2 Direct stimulation of catecholamine receptors is probably not responsible for tyrosine's pressor effect. If tyrosine acted through direct occupancy of catecholamine receptors, its pressor effect would not be blocked by carbidopa, or by hexamethonium, as previously discussed (10). Neither carbidopa nor hexamethonium have been reported to interact with catecholamine receptors.

3.3 Tyrosine's major site of action is the sympathoadrenal system during hypotension. Carbidopa inhibits peripheral amino acid decarboxylase, and does not significantly cross the blood-brain barrier (in 100 mg/kg doses) (10). Phentolamine likewise does not cross the blood-brain barrier; it occupies peripheral catecholamine receptors (10). Hence, the ability of carbidopa or phentolamine to block the tyrosine effects are compatible with a peripheral site of action, as is our previous observation that adrenalectomy also blocks tyrosine's pressor effect (9).

4.0 Clinical implications for tyrosine's effects on BP and catecholamine synthesis during hypotension. Tyrosine might be preferred over traditionally-used pressors during hypotension for several reasons. First, tyrosine can be administered by bolus injection, where norepinephrine, epinephrine, or phenylephrine must be administered by intravenous infusion. (The preparation for intravenous infusions is more costly, cumbersome, and time consuming.) Second, since catecholamine synthesis and release (and therefore tyrosine's pressor effect) are influenced by neuronal feedback, tyrosine's administration does not increase BP above normal. With traditionally used pressors, overshoots in BP may occur easily.

Tyrosine's ability to increase BP during hemorrhage is blocked by concomitant administration of valine, a LNAA that competes with tyrosine for brain uptake (6). The administration or consumption of LNAAs may therefore block the ability of a patient to respond to stress. Hyperalimentation solutions, used for parenteral nutrition, contain only minute quantities of tyrosine when compared with those of other large neutral amino acids (tyrosine is one of the least soluble amino acids) (6), and their effects on a patient's response to hemorrhage has not been studied. Second, a patient's nutritional status, specifically the plasma content of tyrosine, compared with contents of the other LNAAs, may be a useful predictor of neuronal tyrosine and thus is his or her ability to continue to release catecholamines during hemorrhage.

5.0 OTHER CARDIOVASCULAR EFFECTS OF TYROSINE

5.1 Tyrosine decreases BP during hypertension. During hypertension, catecholamine-containing neurons probably from the A anatomic cell group, fire frequently onto brain stem inhibitory neurons, presumably attempting to decrease BP towards normal (24,25). In this situation (hypertension), the neurons that mediate increases in sympathetic activity are quiescent. Since these vasodepressor neurons fire frequently during hypertension, tyrosine might be predicted to similarly decrease BP. In hypertensive animals, tyrosine administration, or consumption of diets with additional tyrosine, decreases BP (26-28). In rats, the BP decreases parallel an increase in brain stem norepinephrine release, suggesting that tyrosine decreases blood pressure by increasing catecholamine synthesis and release in brain stem noradrenergic neurons (26). The simultaneous administration of another LNAA (valine) that competes with tyrosine for entry into the brain, blocks tyrosine's antihypertensive effect (26). In addition, if diets are supplemented with relatively small amounts of tyrosine, the onset of hypertension in spontaneously hypertensive rats is delayed (29). In hypertensive humans, tyrosine decreases BP modestly, with a dramatic effect observed in 1/3 to 1/4 of the patients studied (30). Experiments have not yet been conducted to determine if one subgroup of hypertensive patients is more likely to respond to the amino acid.

5.2 The effects of tyrosine on ventricular arrhythmias. Sudden death syndrome claims as many as 450.000 lives per year (31). Since autopsies frequently do not demonstrate acute myocardial lesions in these patients, the syndrome may represent an electrical accident, such as ventricular

fibrillation (VF), rather than chronic, irreversible myocardial damage (31). CNS signals affect cardiac irritability: for example, one's chances of sudden death are doubled during six months following the death of a spouse, and patients with coronary artery disease also exhibit an increased number of extrasystoles while speaking before a group (31,32). Though these changes may represent ischemia resulting from an increase in BP and heart rate, evidence suggests that sympathetic activity directly affects myocardial vulnerability as well.

Stellate ganglion stimulation induces VF in 60 % of dogs subjected to electrical stimulation of the right ventricle (this experimental model is used to study cardiac vulnerability, 33,34). In the absence of stellate ganglion stimulation, fibrillation does not occur. Likewise, pharmacological manipulations that mimic sympathetic activation decrease the threshold for, or increase the susceptibility to ventricular arrhythmias (35). Norepinephrine infusion (in animals whose BP was held constant by exsanguination) decreases the threshold for VF by almost 50 %. On the other hand, reflex reductions in sympathetic activity from constriction of the thoracic aorta, or from the BP elevation following phenylephrine infusion, increases the VF threshold by 49 % (36). Thus, increases in sympathetic neural activity increase the susceptibility to ventricular arrhythmias, while decreases in sympathetic tone exert a protective effect.

Since treatments that mimic catecholamine release protect the heart from ventricular arrhythmias, Scott et al. tested tyrosine's effect on VF in dogs (37). Tyrosine protected the heart in a dose-related manner. Its protective effect was blocked by the concomitant administration of the large neutral amino acid, valine (37). Though not statistically significant, valine seemed to decrease the threshold when administered alone in the six dogs studied.

6.0 In summary, tyrosine increases BP during hemorrhagic shock by increasing catecholamine synthesis. The amino acid does not interact directly with catecholamine receptors, or increases BP through its conversion to tyramine and may therefore prove useful in the treatment of hemorrhagic shock. Moreover, a patient's nutritional status may be a determinant in his ability to respond to hemorrhage.

REFERENCES

1. Chien, S. Physiol. Rev. 47:214-289,1967.

2. Wurtman, R.J., Larin, F., Mostafapour, S., Fernstrom, J.D. Science 1985:183-184,1974.

3. Wurtman, R.J., Hefti, F., Melamed, E. Pharmac. Rev. 32:315-335,1981.

4. Carlsson, A., Lindqvist, M. Naunyn-Schmiedebergs Arch. exp. Path. Pharamak. 303:157-164,1978.

5. Gibson, C.J., Wurtman, R.J. Life Sci. 22:1399-1406,1978.

6. Melamed, E., Hefti, F., Wurtman, R.J. Proc. natn. Acad. Sci. USA 77: 4305-4309,1980.

7. Sved, A.F., Fernstrom, J.D., Wurtman, R.J. Life Sci. 25:1293-1300, 1979.

8. Conlay, L.A., Maher, T.J., Wurtman, R.J. Spinal cord noradrenergic neurons are activated by hypertension. Am. J. Physiol., submitted for publication.

9. Conlay, L.A., Maher, T.J., Wurtman, R.J. Science 212:559-560,1981.

10. Conlay, L.A., Maher, T.J., Wurtman, R.J. Brain Res. 333:81-84,1985.

11. Conlay, L.A., Maher, T.J., Moses, P.L., Wurtman, R.J. J. neural. Transm. 58:69-74,1983.

12. Conlay, L.A., Zeisel, S.H. Neurosurgery 10:524-529,1982.

13. Conlay, L.A., Maher, T.J. Clin. Anesth. 4:353-361,1984.

14. Pardridge, W.M.,Oldendorf, W.H. J. Neurochem. 28:5-12,1977.

15. Fernstrom, J.D., Wurtman, R.J., Hammarstrom-Wiklund, B., Rand, W.M., Munro, H.N., Davidson, C.S. Am. J. clin. Nutr. 32:1912-1922,1979.

16. Lovenberg, W., Brunswick, E.A., Hanbauer, I. Proc. natn. Acad. Sci. USA 72:2955-2958,1982.

17. Weiner, N., Lee, F.L., Dreyer, E., Barnes, E. Life Sci. 22: 1197-1216,1978.

18. Bowsher, R.R., Henry, D.P. J. Neurochem. 40:992-1002,1983.

19. Tallman, J.F., Saavedra, J.M., Axelrod, J. J. Pharmac. exp. Ther. 199:216-221,1976.

20. Fellman, J.H., Roth, E.S., Fujita, T.S. Archs Biochem. Biophys. 174:562-567,1976.

21. Edwards, D.J., Life Sci. 30:1427-1434,1982.

22. Jurio, A.V., Boulton, A.S. J. Neurochem. 39:859-863,1982.

23. Conlay, L.A., Maher, T.J., Wurtman, R.J. Life Sci. 35:1210-

1212,1984.

24. Struyker, H.A.J., Smeets, G.M.W., Brouwer, G.M., Van Rorrum, J.N. Neuropharmacology 13:837-846,1974.

25. Hausler, G. Circulation Res. 36, Suppl 1:223-232,1975.

26. Sved, A.F., Fernstrom, J.D., Wurtman, R.J. Proc. natn. Acad. Sci. USA 76:3511-3514,1979.

27. Bresnahan, M.R., Hatzinikolaw, A., Brunner, H.R., Gavras, H. Am. J. Physiol. 239:206-210,1980.

28. Osumi, Y., Tanaka, C., Takari, S. Jap. J. Pharmacol. 24:715-720,1974.

29. Bossy, J., Guidoux, R., Milan, H., Wurzner, H.P. Z. ErnährWiss. 22:45-49,1983.

30. Maron, J. personal communication

31. Lown, B., Wolf, M. Circulation 44:130-142,1971.

32. Taggert, P., Parkinson, P., Carruthers, M. Br. med. J. 3:71-76,1972

33. Matta, R.S., Verrier, R.L., Lown, B. Am. J. Physiol. 230:1461-1466,1976.

34. Verrier, R.L., Thompson, P.L., Lown, B. Cardiovasc. Res. 8:602-610,1974.

35. Verrier, R.L., Rabinowitz, S.H., Lown, B. Clin. Res. 23:212A,1975.

36. Verrier, R., Calvert, A., Lown, B. Am. J. Physiol. 226:893-897,1974.

37. Scott, N.A., DeSilva, R.A., Lown, B., Wurtman, R.J. Science 211: 727-729,1980.

6

FURTHER EVIDENCE FOR A VASODEPRESSOR ROLE OF NEUROPEPTIDE Y (NPY) AND ADRENALINE (A) IN THE CENTRAL NERVOUS SYSTEM (CNS) OF THE RAT.

HÄRFSTRAND, A. (1), AGNATI, L.F. (2), FUXE, K. (1), KALIA, M. (3)

(1) Dept. of Histology, Karolinska Institutet, Stockholm, Sweden. (2) Dept. of Human Physiology, Univ. of Modena, Modena, Italy. (3) Dept. of Pharmacology, Thomas Jefferson Medical Center, Philadelphia, USA

ABSTRACT

Previous studies have shown that adrenaline (A), clonidine and neuropeptide Y (NPY) given intracisternally into the α-chloralose anaesthetized male rat produces hypotension, bradycardia and bradyapnea. Only the effects of NPY are resistant to blockade of α-2 adrenergic receptors. However, the NPY and α-2 adrenergic receptors probably interact since no additive effects are observed on cardiovascular parameters following coadministration of various doses of clonidine and NPY or adrenaline and NPY. Furthermore, analysis of serum catecholamine levels shows a selective reduction of the noradrenaline levels.
Comparisons of central administration of NPY and A into the awake freely moving rat versus the α-chloralose anaesthetized rat show that with maximal doses of the two compounds the cardiovascular responses are reduced in the awake freely moving rats. However, the vasodepressor area seen after centrally administered NPY is larger than that observed after centrally administered A using maximal doses.
The possible underlying mechanisms and site of action are discussed.

INTRODUCTION

NPY-like immunoreactivity has been demonstrated in a vast majority of A nerve cell bodies in the medulla oblongata of the rat (1-3). Recently indications have also been obtained that NPY-like immunoreactivity may exist in a large subpopulation of the dorsal strip (ds) adrenergic neurons (4). A large degree of coexistence of phenylethanolamine-N-methyltransferase (PNMT) (used as an immunocytochemical marker for A neurons) and NPY-like immunoreactivity has also been observed in nerve terminals of the dorsal motor nucleus of the vagus (dmnX) and of the medial subnucleus of the nucleus tractus solitarius (mnTS), (5). However, it must be underlined that NPY-like immunoreactivity is also present in other types of neurons some of which contain tyrosine-hydroxylase (TH) and dopamine ß-hydroxylase (DBH) immunoreactivity and therefore probably represent noradrenergic neurons (especially group A1 but also parts of group A2 of the medulla oblongata). Nevertheless, it seems likely that at least a part of the vasodepressor activity exerted by NPY is related to its role as a co-transmitter in a large number of adrenergic synapses expecially in those of the nTS and of the sympathetic lateral column (5-7). The observations that NPY could reduce arterial blood pressure (ABP), heart rate (HR) and respiration rate (RR) upon central administration (6) was of particular interest since we believe that many adrenergic neurons have an important vasodepressor function and that clonidine can induce its blood pressure lowering action via activation of α_2 adrenergic receptors linked to A synapses in cardiovascular centra of the medulla oblongata and the spinal cord and also in the hypothalamus (8-11). The vasodepressor actions of NPY and the NPY related peptide peptide YY (PYY) have been further explored in the present paper as well as the interaction of NPY with the α_2 adrenergic receptors. It was previously shown that the hypotensive action of NPY was not prevented by pretreatment with an α_2 adrenergic antagonist idazoxan (7) given intracisternally (i.c.) 20 minutes prior to the injection of NPY. However, the action of clonidine also given i.c. was completely abolished after pretreatment with idazoxan. The present studies are concerned with the further analysis of the role of the NPY receptors in the control of central cardiovascular functions as well as with its interaction with the α_2 adrenergic receptors, using intraventricular (i.v.t) and i.c. injections of A, NPY and clonidine in α-chloralose anaesthetized and awake unrestrained male rats

MATERIAL AND METHODS

Specific pathogen free male Sprague-Dawley rats (200-300 g body weight), (ALAB, Stockholm) were used. Physiological experiments were performed both in α-chloralose anaesthetized rats (100 mg/kg i.v. lingual vein) as well as in awake unrestrained freely moving rats. In the experiments on the anaesthetized animals i.c. injections (10 ul) of A, clonidine, idazoxan, NPY and PYY were performed stereotaxically while in the experiments on awake animals i.v.t. injections (lateral ventricle) were performed by means of a microinjection unit, (30 ul mock CSF was used as a solvent, pH 7.5). Infusion time was 3 minutes. In the experiments on the awake rats the cannula were implanted into the lateral ventricle 2 days before the experiment by means of a stereotaxic instrument (12). Mean ABP and HR were recorded by means of a heparinized catheter (50 I.U/ml saline) in the common carotide artery using a Statham P23 transducer connected to a Grass polygraph (model 7). Respiratory frequency was simultaneously recorded via an intraoesophageal catheter (also linked to the polygraph via a transducer) or by ocular inspection (awake animals). Basal levels of HR, RR and ABP were recorded for a 30 minute period prior to the injection of the various drugs. The rats were analyzed for 60 to 120 minutes following the injection of the peptides, A or clonidine. The body temperature of the anaesthetized rats was kept at $37° \pm 0.5°$ C by an automatic heating device regulated via a rectal probe.

NPY and PYY were obtained from Bachem, Bubendorf, Switzerland. A was obtained from Sigma, St Louis, USA. In some experiments the highly selective α_2 adrenergic antagonist idazoxan (RX781094) was used (generous gift from Reckitt and Coleman, Hull U.K.). Clonidine was generously supplied by Boehringer-Ingelheim, FRG.

RESULTS

Cardiovascular effects of clonidine and A

In Fig. 1 it is shown that A and clonidine induce a dose dependent lowering of mean arterial blood pressure (MAP). Also a dose related lowering of HR is observed with both A and clonidine given i.c. into the α-chloralose anaesthetized male rat. The peak action is reached for both A and clonidine in a dose around 1 nmol. Also in the awake unrestrained male rat A is shown to produce a reduction of ABP (Fig. 2). However, the effects are less pronounced than those seen following administration of the same amount of A into the anaesthetized rat. The dose used in the experiments on awake animals, illustrated in Fig. 4, is the dose that produced a maximal hypotensive response.

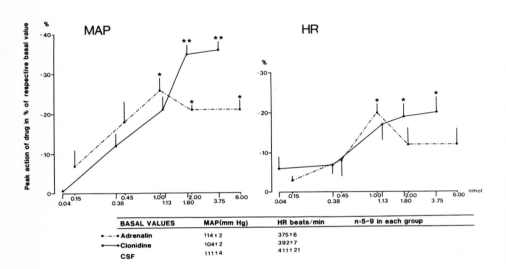

CARDIOVASCULAR EFFECTS OF INTRACISTERNAL ADMINISTRATION OF CLONIDIN AND ADRENALINE
IN THE α CHLORALOSE ANETHETIZED MALE RAT

BASAL VALUES	MAP(mm Hg)	HR beats/min	n=5-9 in each group
Adrenalin	114 ± 2	375 ± 6	
Clonidine	104 ± 2	392 ± 7	
CSF	111 ± 4	411 ± 21	

Fig. 1   Effects of i.c. administration of clonidine and A in the
α-chloralose anaesthitized male rat are shown on mean arterial
blood pressure (MAP) and heart rate (HR). The dose is expressed
in nmol/rat. The drugs were dissolved in mock CSF and the injection
volume was 10 ul. The effects shown are the peak effects (0-60
minute interval) taken in % of respective basal value after
subtraction of the slight effect of mock CSF alone. Means ± s.e.m.
are shown. Statistical analysis was carried out by means of
Wilcoxon test (nonparametrical) treatment versus control.
* = $p < 0.05$; ** = $p < 0.01$

Cardiovascular effects of centrally administered NPY

In the α-chloralose anaesthetized male rat it is found that NPY given i.c. produces a dose related lowering of MAP and of HR (see Fig. 2). If anything NPY was more potent than A (see Fig. 1). The maximal effects observed with NPY were similar to those seen with A (see Fig. 1). Also in the awake unrestrained male rat NPY produces a marked and polonged lowering of ABP, HR and respiration rate (Fig. 3). When comparing the effects of NPY with the effects of A after central administration in the awake and α-chloralose anaesthetized male rat it is shown that NPY produces a similar peak action as A but the vasodepressor area during the 60 minutes analyzed is significantly larger following administration NPY compared with that observed after administration of A in the awake unrestrained male rat. Like A, NPY produces a more marked lowering of ABP in the α-chloralose anaesthetized male rat compared with the awake conscious male rat (see Fig. 4). However, the bradycardic area obtained following administration of NPY in the anaesthetized or the awake male rat is not significantly different from the that obtained following the injection of A. In Fig. 5 it is shown that the i.c. injection of NPY in the α-chloralose anaesthetized male rat leads to a significant lowering of serum noradrenaline (NA) levels at the time of the peak action.

Failure of an α_2 adrenergic antagonist idazoxan to reverse and to prevent the cardiovascular actions of NPY

In these experiments clonidine was included for comparison (3.7 nmol given i.c. in 10 ul mock CSF). It is shown in Fig. 6 that idazoxan can efficiently counteract the hypotensive, bradycardic and bradypneic actions of clonidine but cannot reverse the cardiovascular and respiratory effects of NPY (1.25 nmol). Similar results were obtained in the prevention experiments (Fig. 7). Thus, idazoxan (205 nmol i.c.) was capable of preventing the hypotensive actions and the bradycardic actions of clonidine but not of NPY. Similar results have also been obtained with PYY (3). These pharmacological experiments were performed in the α-chloralose anaesthetized male rat.

Cardiovascular effects of combined treatment with NPY/clonidine and NPY/A

It is shown that no additive effects are observed, when clonidine and NPY are given together in subtreshold, submaximal (around ED_{50} value) or

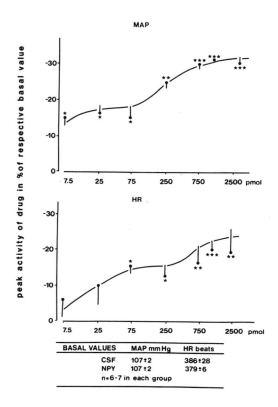

EFFECTS OF INTRACISTERNAL ADMINISTRATION OF NPY
IN THEα-CHLORALOSE ANESTHETIZED MALE RAT.

Fig. 2 Effects of i.c. administration of NPY in the α-chloralose
anaesthetized male rat on MAP and HR. Administered doses are shown
as pmol/rat. The drugs were dissolved in mock CSF and the injection
volume was 10 ul. The effects shown are the peak effects (time
interval 0-60 minutes) taken in % of respective basal value after
subtracting the slight effect of mock CSF alone. Means + s.e.m.
are shown. Statistical analysis was carried out by means of
Wilcoxon test (non-parametrical), treatment versus control.
* =p<0.05; ** = p<0.01; *** = p<0.001

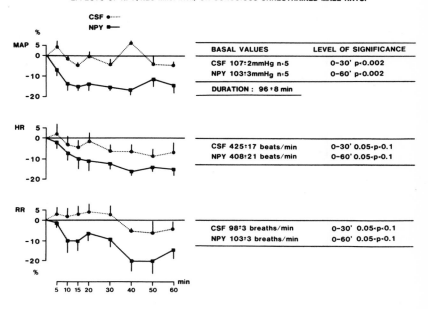

EFFECTS OF NPY(1.25 nmol i.v.t.) ON CONSCIOUS UNRESTRAINED MALE RATS.

BASAL VALUES	LEVEL OF SIGNIFICANCE
CSF 107±2mmHg n:5	0-30' p<0.002
NPY 103±3mmHg n:5	0-60' p<0.002
DURATION : 96±8 min	

BASAL VALUES	LEVEL OF SIGNIFICANCE
CSF 425±17 beats/min	0-30' 0.05<p<0.1
NPY 408±21 beats/min	0-60' 0.05<p<0.1

BASAL VALUES	LEVEL OF SIGNIFICANCE
CSF 98±3 breaths/min	0-30' 0.05<p<0.1
NPY 103±3 breaths/min	0-60' 0.05<p<0.1

Fig. 3 Cardiovascular and respiratory effects of NPY after i.v.t. administration in the conscious unrestrained freely moving male rat. The effects of a high dose of NPY (1.25 nmol, ceiling effect) are shown. The rats have been cannulated 2 days earlier. Mock CSF was used as solvent and injection volume was 30 ul injected during 3 minutes. The time course of the effects of NPY is shown for the first 60 minutes. Means ± s.e.m. are given for the CSF group and the NPY treated group. The duration of the hypotensive action of NPY was 96 ± 8 minutes. For statistical analysis the hypotensive, bradycardic and bradypneic area under the curve (AUC) were analyzed both for the CSF and NPY group and then compared using the Mann-Whitney U-test (nonparametrical). After treatment with NPY a highly significant lowering of ABP was demonstrated as well as a trend for a reduction of HR and RR.

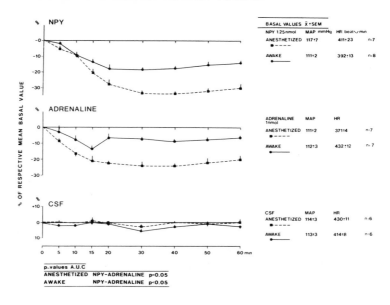

COMPARISONS OF THE EFFECTS OF A AND NPY ON MAP AFTER CENTRAL ADMINISTRATION IN THE AWAKE (IVT) AND α-CHLORALOSE ANESTHETIZED (IC) MALE RAT

Fig. 4   The effects of ceiling doses of A and NPY on MAP are compared after
i.c. administration in the α-chloralose anaesthetized male rat or
after i.v.t. injections in the awake unrestrained freely moving
male rat. NPY was administered in a dose of 1.25 nmol/rat dissolved
in mock CSF. The injection volume was 10 ul for i.c. injections
and 30 ul for i.v.t. administration. The dose of A was 1 nmol/rat.
The results are given in % of respective basal values. Means ±
s.e.m. are shown. The NPY and A vasodepressor actions are shown to
have similar duration but a shorter latency to the peak after A
administration in the awake animals. In both NPY and A treated
animals the effects were more marked in the anaesthetized animals.
The hypotensive actions of centrally administered NPY was
significantly larger in both anesthetized and unanaesthetized
animals than that of A. The peak action of NPY in the awake
animals is prolonged compared with A treated animals. For
statistical analysis all data were computerized and the area under
the curve value was calculated for each parameter and compared
using the Mann-Whitney U-test (non-parametrical).

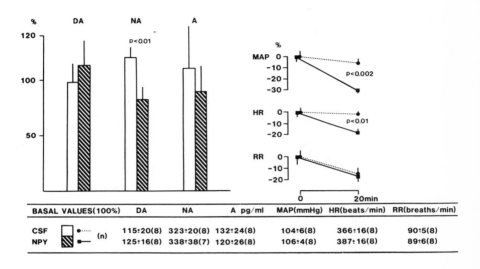

EFFECTS OF NPY(1.25 nmol i.c.)ON SERUM CATECHOLAMINE LEVELS IN THE α-CHLORALOSE ANAESTHETIZED MALE RAT

BASAL VALUES(100%)		DA	NA	A pg/ml	MAP(mmHg)	HR(beats/min)	RR(breaths/min)
CSF	●····· (n)	115±20(8)	323±20(8)	132±24(8)	104±6(8)	366±16(8)	90±5(8)
NPY	■—	125±16(8)	338±38(7)	120±26(8)	106±4(8)	387±16(8)	89±6(8)

Fig. 5   The effects of intracisternal administration of a high dose of NPY (1.25 nmol) on serum catecholamine levels in the α-chloralose anaesthetized male rat. To reveal the mechanism behind the hypotensive action of NPY after central administration arterial blood (300 ul) was collected before NPY or CSF treatment and at the time of peak of the hypotensive action (20 - 25 minutes). Means ± s.e.m. are shown. Noradrenaline (NA) but not A levels was significantly reduced in the NPY treated animals at the time of the peak action. Mann-Whitney U-test was used (nonparametrical) for statistical analysis. The catecholamines were analyzed by means of a RIA procedure. (Upjohn CAT-A-KIT)

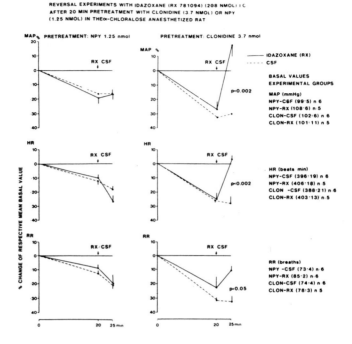

REVERSAL EXPERIMENTS WITH IDAZOXANE (RX 781094) (208 NMOL) I.C.
AFTER 20 MIN PRETREATMENT WITH CLONIDINE (3.7 NMOL) OR NPY
(1.25 NMOL) IN THEα-CHLORALOSE ANAESTHETIZED RAT

Fig. 6 The highly selective α_2 adrenergic antagonist idazoxan (RX781094)
was used in attempts to reverse cardiovascular changes induced
by NPY and clonidine. NPY or clonidine was administered 20 minutes
prior to idazoxan. Means $\pm$ s.e.m. are shown in % of respective
basal value at 20 minutes (just before clonidine/NPY
administration) and at 25 minutes (5 minutes before clonidine/NPY
administration). In the clonidine treated animal idazoxan produces
a rapid reversal of the hypotensive, bradycardic and bradypneic
effect, while the NPY treated animals are unaffected and maintain
their hypotension, bradycardia and bradypnea. Significant
differences between the drug treated and mock CSF treated groups
were revealed by the use of Mann-Whitney U-test (nonparametrical).

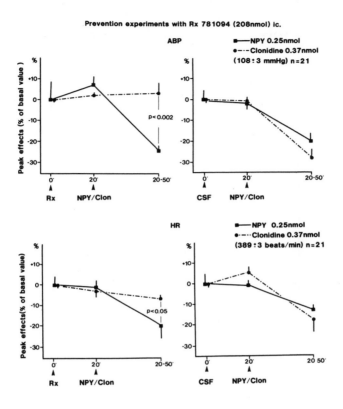

Fig. 7   Effects of i.c. injections of the highly selective α_2 adrenergic
receptor antagonist idazoxan (RX781094) on the hypotensive and
bradycardic action of i.c. injections of clonidine and NPY in the
α -chloralose anaesthetized male rat (prevention experiment).
Idazoxan was administered 20 minutes prior to injection of NPY or
clonidine. Data are shown in % of respective basal value (means $\pm$
s.e.m.). The effect of clonidine and NPY after pretreatment with
idazoxan are shown as the peak value (20 - 25 minutes time
interval). Mann-Whitney U-Test (nonparametrical) was used for the
statistical analysis.

maximal (ceiling effect) doses into the cisterna magna of the α-chloralose anaesthetized male rat (Fig. 8) compared with the hypotensive, bradycardic and bradypneic actions of NPY and clonidine alone in the same doses. Similar results were observed when A and NPY were given together i.c. in submaximal doses (Fig. 9).

DISCUSSION

The present results give evidence that centrally administered NPY and the related pancreatic polypeptide PYY reduces ABP, HR and RR probably via an action on NPY recognition sites in the CNS. Receptor autoradiographical studies on iodinated NPY binding sites reveal high densities in the nTS (13) indicating that one major site of action may be the nTS area. Another site of action may be the sympathetic lateral column since this nucleus is richly innervated by NPY immunoreactive nerve terminals. When discussing the nTS as a site of action for the vasodepressor activity of NPY and A it should be considered that the ds region of the nTS where the baroreceptors terminate contain numerous small A nerve cell bodies and also numerous small NPY neurons many of which contain both NPY and PNMT-like immunoreactivity (13). The ds neurons probably mainly respresent local neurons innervating other regions of the nTS including the medial subnucleus, where a large number of projection neurons appear to exist, including NA cell bodies of group A2. A few of these neurons appear to costore NPY-like immunoreactivity at the level of the rostral part of the area postrema. It is therefore postulated that A and NPY given i.c. reduce ABP and HR at least in part by activating the baroreceptor reflex pathway within the nTS. Thus, NPY/A costoring neurons of the ds are activated by baroreceptor reflex afferents to inhibit activity in outgoing projection neurons with vasodepressor activity such as possibly the noradrenergic A2 neurons which innervate the spinal cord and the hypothalamus (14).

Another important site of action may be the sympathetic lateral column richly innervated by NPY and PNMT immunoreactive nerve terminals. Many of these nerve terminals may costore PNMT and NPY-like immunoreactivity, since they are predominantly innervated by axons, originating from group C1 in the medulla oblongata (rostral and ventrolateral parts). Most C1 A cells are known to contain both NPY and PNMT-like immunoreactivity (1, 2). The sympathetic lateral column is also known to contain a high density of α_2

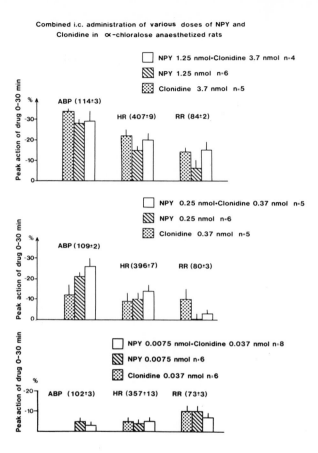

Fig. 8 Effects of combined i.c. administration of various doses of NPY and clonidine on ABP, HR and RR compared with the substances alone in the α-chloralose anaesthetized male rat. Peak effect of the drugs are shown during the first 30 minutes following injection and are expressed in % of respective basal mean value (means ± s.e.m.). The drugs were dissolved in mock CSF, and the effects shown represent the effects observed after subtracting the slight effects produced by mock CSF alone on each parameter. No statistical differences were revealed between any of the groups using Wilcoxon test (nonparametrical).

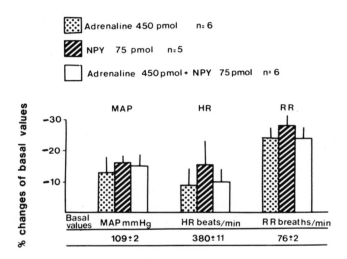

COMBINED I.C. ADMINISTRATION OF SUBMAXIMAL DOSES ADRENALINE AND NPY INαCHLORALOSE ANESTHETIZED RATS

Fig. 9   Effects of combined i.c. administration of submaximal doses
(around ED 50 valvue) of NPY and A on ABP, HR and RR in the
α-chloralose anaesthetized male rat. The peak effects of the
drugs are shown during the first 30 minutes following injection
and are expressed in % of respective basal mean value (means +
s.e.m.). No statistical differences were revealed between any of
the groups using Wilcoxon test (nonparametrical).

adrenergic receptors (15).

An interesting finding in the present paper is the failure of additive effects on ABP and HR, when A and NPY or NPY and clonidine were coadministered into the cisterna magna in the α-chloralose anaesthetized rat. One mechanism underlying the failure of additive actions may be that both the α_2 adrenergic receptors and the NPY receptors are coupled to the same Ni protein since both α_2 agonists and NPY are known to inhibit adenylate cyclase activity by activating the GTP binding Ni protein (16,17). In line with this interpretation it has been shown in in vitro experiments that NPY can modulate the binding characteristics of α_2 adrenergic agonist and antagonist binding sites (18,19). Receptor autoradiographical studies also show that NPY can modulte 3H-para-aminoclonidine binding (α_2 agonist binding) in the nTS (5). An additional mechanism may be that both at the level of the sympathetic lateral column and of the nTS NPY and A act on the same final common pathway for the vasodepressor activity. Furthermore, NPY and A given i.c. may switch off the activity in the A and NPY transmission lines respectively (5,13). It should also be considered that at the molecular level activation of e. g. NPY receptors may lead to the formation of a cluster with adjacent α_2 adrenergic receptors via receptor-receptor interactions. Such an interaction may also have an inhibitory function. All these mechanisms should be considered together when understanding the interactions between adrenergic and NPY mechanisms in central cardiovascular regulation and in their failure to demonstrate additive actions when administered together.

In conclusion the present results open up the possibility NPY-like molecules specific for brain NPY receptors may offer a complementary treatment of essential hypertension. This assumption gains support from the observations that NPY induces a more prolonged reduction of MAP in the awake unrestrained male rat compared with A itself and also in the fact that in the SH rat does not induce sedation but arousal, seen as an increase in EEG desynchronization (20).

ACKNOWLEDGEMENT

This work has been supported by a grant from the Swedish Medical Research Council, by a grant from Kurt and Alice Wallenbergs Stiftelse, and

108

by a grant from The Swedish Society of Medical Sciences. For excellent technical assistance we are grateful to Lars Rosen. For excellent secreterial assistance we are grateful to Ms Anne Edgren and Ms Rose Marie Gustafsson.

REFERENCES

1. Hökfelt, T., Lundberg, J.M., Tatemoto, K., Mutt, V., Terenius, L., Polak, J., Bloom, S., Elde, R., Goldstein, M. Acta physiol. scand. 117: 315-318, 1983.
2. Everitt, B.J., Hökfelt, T., Terenius, L., Tatemoto, K., Mutt, V. Goldstein, M. Neuroscience Vol. 11, No. 2: pp. 443-462, 1984.
3. Härfstrand, A. et al. in preparation.
4. Härfstrand, A., Kalia, M., Fuxe, K., Terenius, L., Goldstein, M. J. comp. Neurol.: submitted, 1986.
5. Fuxe, K., Agnati, L.F., Härfstrand, A., Janson, A.M., Neymeuer, A., Andersson, K., Ruggeri, M., Zoli, M., Goldstein, M. Prog. Brain Res.: in press, 1986.
6. Fuxe, K., Agnati, L.F., Härfstrand, A., Zini, I., Tatemoto, K., Merlo Pich, E., Hökfelt, T., Mutt, V., Terenius, L. Acta physiol. scand. 118: 189-192, 1983.
7. Härfstrand, A., Fuxe, K., Agnati, L.F., Ganten, D., Eneroth, P., Tatemoto, K., Mutt, V. Clin. exp. Hypertens. A6(10-11):1947-1950, 1984.
8. Fuxe, K., Hökfelt, T., Bolme, P., Goldstein, M., Johansson, O., Jonsson, G., Lidbrink, P., Ljungdahl, A., Sachs, CH. In: Central Action of Drugs in Blood Pressure Regulation (Eds. D.S. Davies, J.S. Reid), Pitman Medical London, pp. 8-22, 1975.
9. Bolme, P., Corrodi, H., Fuxe, K., Hökfelt, T., Lidbrink, P., Goldstein, M. Eur. J. Pharmacol. 28: 89-94, 1974.
10. Fuxe, K., Bolme, P., Jonsson, G., Agnati, L.F., Goldstein, M., Hökfelt, T., Schwarcz, R., Engel, J. In: Nervous Systems and Hypertension, Wiley-Flammarion, Paris, pp. 1-17, 1979.
11. Franz, D.N., Madsen, P.W. Eur. J. Pharmacol. 78: 53-59, 1982.
12. Härfstrand, A., Fuxe, K., Agnati, L.F., Zini, I., Zoli, M., Andersson, K., Eneroth, P., von Euler, G., Terenius, P., Mutt, V.,

Goldstein, M. Neurochem. Int.: in press, 1985.

13. Fuxe, K. et al. this symposium.

14. Kalia, M. In: International Symposium on Brain Epinephrine, Sept. 29
    - Oct. 3, Baltimore, U.S.A., (Eds. J.M. Stolk, D. U'Prichard
    & K. Fuxe), 1985.

15. Unnerstall, J.R., Palacios, J.M., Kuhar, J.M. Soc. Neurosci. 7: 501,
    1981.

16. Jakobs, K.H., Aktories, K., Schultz, G. Adv. Cyclic Nucleotide Res.
    14: 173-187, 1981.

17. Fredholm, B.B., Jansen, I., Edvinsson, L. Acta physiol. scand. 124:
    467-469, 1985.

18. Agnati, L.F., Fuxe, K., Benfenati, F., Battistini, N., Härfstrand, A.,
    Tatemoto, K., Hökfelt, T., Mutt, V. Acta physiol. scand. 118: 293-295,
    1983.

19. Fuxe, K., Agnati, L.F., Härfstrand, A., Martire, M., Goldstein, M.,
    Grimaldi, R., Bernardi, P., Zini, I., Tatemoto, K., Mutt, V. Clin.
    exp. Hypertens. A6(10&11): 1951-1956, 1984.

20. Zini, L., Merlo Pich E., Fuxe, K., Lenzi, P.L., Agnati, L.F.,
    Härfstrand, A., Mutt, V., Tatemoto, K., Moscara, M. Acta physiol.
    scand. 122: 71-77, 1984.

7

THE RENIN-ANGIOTENSIN-SYSTEM IN THE BRAIN

GANTEN, D., UNGER, Th., LANG, R.E.

German Institute for High Blood Pressure Research and Department of Pharmacology, University of Heidelberg, Im Neuenheimer Feld 366, D-6900 Heidelberg, FRG

ABSTRACT

The biosynthetic pathway and the functions of the hormonal plasma renin angiotensin system are well recognized. Since the components of the RAS were first described within the brain (Ganten et al., Science 1973:64, 1971), evidence has accumulated supporting the existence of a complete RAS endogenous to the brain and playing a role in cardiovascular and volume homeostasis. The high molecular weight peptide precursor angiotensinogen has been cloned and the complete sequence of angiotensinogen is determined. Using cellfree translation techniques it has been demonstrated that the same angiotensinogen molecule is synthesized in the brain and in the liver. The regulation of transcription of the gene, however, appears to be tissue specific. Angiotensin I and the active octapeptide angiotensin II (ANG II) have been characterized in the brain. The distribution of ANG II in the central nervous system has been investigated by chemical as well as by immunohistochemical techniques. The main locations of ANG II are the hypothalamus, the limbic system, the medulla oblongata and the spinal cord; there is good agreement with the localization of ANG II receptors. The key enzymes of the peptide system, i.e. renin, converting enzyme and angiotensinases have equally been shown to be present in the brain and pharmacological inhibition of the synthesizing enzymes leads to reduced angiotensin synthesis and e.g. blood pressure reduction (Ganten et al., Science 221: 869, 1983). Since the genes for the RAS proteins are now known, studies at the cellular and molecular level have become possible. The availability of pharmacological interferences at the sites of peptide generation and peptide action to an extent not equalled for other enzyme peptide systems makes the RAS one of the best studied peptide systems in the brain.

Angiotensin II (ANG II) is well known as a circulating peptide hormone. Plasma angiotensin exerts its blood pressure effects mainly by the stimulation of aldosterone secretion and by the enhancement of vasoconstriction. Plasma angiotensin II also effects the brain at sites were the blood brain barrier is deficient and induces blood pressure increases, drinking and the release of pituitary hormones; but these central effects of angiotensin are less pronounced compared to ANG II given directly into the brain. Indeed, there is now convincing evidence that ANG II in addition to being a plasma hormone also occurs locally in the brain as a neuropeptide. The evidence that brain ANG II as well as plasma ANG II are involved in blood pressure control and the maintenance of hypertension comes mainly from experimental data.

In stroke-prone spontaneously hypertensive rats (SHRSP) elevated levels of renin were found in catecholaminergic nuclei of the brain, in the pineal organ and in the adeno- and neurohypophysis (1). Angiotensin synthesis and degradation ("turnover") are higher in young SHRSP (2); CSF angiotensinogen was found to be elevated in hypertensive patients, suggesting a stimulated brain RAS in hypertensive subjects (3). Progesterone treatment resulted in an elevation of brain renin in dogs (4).

Pharmacological interference with the brain RAS at various levels of the enzyme-peptide cascade has been used to investigate whether locally generated endogenous brain angiotensin was involved in blood pressure regulation (5-8).

Acute blockade of brain angiotensin receptors by intraventricular (i.c.v.) administration of competitive ANG II antagonists in SHR led to a consistent, dose dependent decrease of blood pressure (5,9). When kidney renin was eliminated in SHRSP by nephrectomy, the blood pressure decrease was still found after i.c.v. ANG II receptor blockade.

Inhibtion of converting enzyme in the brain with captopril equally lowered blood pressure in SHR. Low doses of i.c.v. captopril (5 ug) caused converting enzyme inhibition in the CSF and in the brain but not in the peripheral blood. Thus, the fall of blood pressure was centrally mediated (7,10).

This observation has been confirmed in studies by Berecek et al. (11) who administered low doses of CE inhibitors chronically to SHR and found the drug to lower blood pressure when administered centrally but not when given by the intravenous route at the same dose. The results indicated that an increase of baroreceptor reflex activity and a blunting of vascular reactivity might be involved in the depressor effect of central CE inhibitors. These data further support a contribution of brain angiotensin in the maintenance of high blood pressure. As reviewed elsewhere (8) there is increasing evidence that orally given CE inhibitors do have central effects and it is thus possible that part of the blood pressure lowering action of CE inhibitors is mediated by an effect of the orally administered drug on the brain.

These functional data made it necessary to investigate the question whether ANG II can be synthetized locally in the brain.

Evidence for the presence of components of the renin angiotensin system (RAS) in the brain of various species has recently been reviewed (6,12-14). We discuss here some recent advances concerning the characterization of renin, angiotensinogen, converting enzyme and angiotensin in the brain and their localization in areas which are relevant to volume homeostasis and cardiovascular control.

The primary structure of renin has been determined by classical amino acid sequence analysis (15) and was derived from the nucleotide sequences of cloned cDNAs complementary to their mRNA (16,17). The mature renin molecule consists of two chains. The heavy chain contains 288, and the light chain 48 amino acid residues. Data from different laboratories are in good agreement. Mouse submaxillary gland renin exhibits a 43 % sequence homology with porcine pepsin, 34 % identity with bovine chymosin and 22 % identity with pencillopepsin (15). The overall dimension and shape of the renin molecule appears to be similar to other acid proteases. Renins isolated from various sources including hog, rat and human kidney renin, bovine pituitary renin, and mouse submaxillary renin are all similar in such general molecular properties as amino acid composition, chain length, molecular weight, isoelectric points (see 13,15-20).

The fact that a number of acid proteases, including cathepsin D, can generate ANG I from angiotensinogen, has lead to a controversy whet er the brain contained "true" renin. This controversy was terminated when it was demonstrated that brain renin is active at neutral pH, can be separated from cathepsin D and other acid proteases, is inhibited by specific renin antibodies and peptide inhibitors, and is active in vivo (13,18-20).

The ultimate proof for local synthesis of renin in the central nervous system will stem from recombinat cDNA techniques as has been shown for brain angiotensinogen. We have used a renin cDNA probe for hybridization studies with mRNA isolated from mouse tissue. The probe, plasmid pMSR 49 (17), contains a 700 bp insert of mouse submaxillary cDNA cloned into pBR 322. This approach was used for hybridization studies with RNA from brain and other organs of male mice (21).

Briefly, Messenger RNA (mRNA) was isolated from total RNA by chromatography on oligo-(dT) cellulose, denatured and then fractioned electrophoretically by size on formaldehyde-agarose gels. The mRNA was then blotted from the gel onto nitrocellulose. Hybridization on these filters was performed with nicktranslate heat denatured (alpha^{32}p)-labeled plasmid pMSR 49 containing submaxillary gland renin cDNA (Northern blot). After washing hybrid bands were visualized by autoradiography using Kodak X-ray screen. Preliminary results show that the hybridization bands in brain corresponded to those found in submaxillary gland and in kidney.

Relative amounts of renin mRNA in mouse organs were determined by dotting analysis. For this purpose mRNA was spotted and fixed on nitrocellulose filters. Hybridization was done under the same conditions as for Northern blotting and the dried filters were counted directly in liquid scintillation fluid or exposed to X-ray film. Direct counting and densitometric analysis of the dot blots revealed the following rank order of hybridization: submaxillary gland $>$ kidney $>$ brain (21).

These data are in harmony with the idea that the renin gene is expressed and that the protein is synthetized locally in brain. However, the final proof will require one or more of various approaches including cell free translation of brain mRNA and identification of the newly synthetized protein as renin, further characterization of the specificity of hybridization by digestion with nuclease SI ("SI mapping"), or sequencing of cDNA transcripts synthetized by reverse transcriptase. Such work is currently underway in several laboratories.

The complete sequence of angiotensinogen was recently determined by recombinant cDNA techniques from a clone selected from a rat liver cDNA bank and subjected to nucleotide sequence analysis (22). The deduced amino acid sequence indicated that the precursor molecule consists of a mature angiotensinogen and a putative signal peptide of 24 amino acids. The predicted molecular weight and amino acid composition of angiotensinogen agreed well with those obtained by amino acid analysis of the purified protein (23). The ANG I moiety is located at the amino terminal part of the molecule, followed by a large carboxy-terminal sequence. This carboxy-terminal sequence contains two small internally homologous sequences and three potential glycosylation sites. The possibility that the carboxyl-terminal region of angiotensinogen has some biological role after the release of ANG I, still awaits investigation.

Using the technique of cell free translation of mRNA, Campell et al. (24) recently provided evidence that the same angiotensinogen molecule is synthetized in the liver and locally in the brain. In their studies, ^{35}S-Methionine labeled angiotensinogen precursors were synthetized by cell free translation of either rat brain or rat liver mRNA and compared by immunoprecipitation, sodium dodecyl sulfate polyacrylamide gel electrophoresis, and autoradiography. Rat liver mRNA synthetized two angiotensinogen precursors: a major precursor of molecular weight 52.5k and a minor precursor of MW 55.7k. Indentical and similarly abundant precursor forms to those observed for liver, were synthetized by cell free translation of rat brain mRNA. Both brain angiotensinogen precursors were cleaved by renin, resulting in a single cleavage product with a molecular weight of 47.5k, identical to that observed for liver. Bilateral nephrectomy and dexamethasone administration produced less than two-fold increase in translatable levels of brain angiotensinogen mRNA, in contrast to the several-fold increase observed for liver. These results show that although rat brain and liver angiotensinogen mRNAs appear to be products of the same gene(s), the regulation of their transciption is tissue specific (24).

CE is widely distributed throughout the brain. The localizations obtained in microdissection studies measuring CE catalytic activity and those described using immunohistochemical techniques are in reasonably good agreement. Recent data were obtained by autoradiographic visualization of CE with ^{3}H-captopril (25).

The highest concentrations occur in the choroid plexus, subfornical organ, caudate-putamen, zona reticulata, substantia nigra, globus pallidus and median eminence. In certain areas (e.g. entopeduncular nucleus, medial habenula, median preoptic area), however, there is disagreement between the autoradiographic and biochemical or immunological data (26). Brain blood vessels contain CE, but the enzyme is clearly also located neuronally as evidenced in cell culture and by ultracentrifugation studies. The latter techniques produced evidence that CE is present in synaptosomes (27-29).

The striato-nigral localization of CE is of particular interest. Ibotenic acid, selectively destroys neuronal cell bodies intrinsic to the site of injection without damaging glial elements, extrinsic nerve terminals and axons of passage (25). Injections of ibotenic acid into the caudate-putamen produced a decrease of CE at the site of injection and, later on, a depletion in the substantia nigra. On the other hand, the same injections into the substantia nigra were without effect on CE activity (25,26). The findings show that CE has a neuronal localization within the corpus striatum and that the CE producing neurons (cell bodies) project to the ipsilateral substantia nigra. Glia seems to be devoid of CE. The decrease of CE in the candate-putamen was associated with an increase in renin activity which could represent a compensatory effect (25). The finding that the typical destruction of the corpus striatum in Huntington's disease is paralleled by a depletion of CE activity in the substantia nigra is noteworthy in this respect (30). Surprisingly it has not been possible to demonstrate ANG II-receptors and ANG II immunoreactivity in the striato-nigral structures. CE may nevertheless play an important role in these structures since the enzyme has a broad specifity and may hydrolyse other peptides as well.

Brain areas where ANG II has been shown to occur with no CE include parts of the spinal cord, the bed nucleus of the stria terminalis and the central nucleus of the amygdala. The significance of this remains to be investigated. The different ratio of ANG I/ANG II in various brain areas (2) would be consistent with the interpretation, that CE determines in certain areas the activity of the RAS in the brain.

The presence of CE in specific brain regions has become of particular interest since inhibitors of the enzyme have been introduced as antihypertensive drugs (8). These agents provide a new tool to study brain peptide metabolism and there is increasing evidence that they act on brain CE and thereby effect volume homeostasis and cardiovascular control even if given orally (7,8,10,31).

Angiotensin I (ANG I) and Angiotensin II (ANG II) have been extracted from brain of nephrectomized rats, rabbits and primates (2,32). Peptides were characterized with high performance liquid chromatography (HPLC) capable of separating all angiotensins and their fragments including (Val[5])-ANG II from (Ile[5])-ANG II. The peptides extracted from brain corresponded to synthetic (Ile^5)-ANG II and (Ile^5)-ANG I, with small amounts (approximately 10 %) of (Ile^5)-ANG(2-8) (ANG III) being present in the brain. Brain angiotensin thus appears to have the same amino acid sequence as plasma angiotensin. The identical peptide was cleaved from brain and plasma angiotensinogen upon incubation with renin in vitro and in vivo (2).

The distribution of ANG II in the central nervous system has been investigated biochemically (2) as well as by immunohistochemical techniques (33-38). The main locations of ANG II are the hypothalamus, the limbic system, the medulla oblongta and the spinal cord. High density of ANG II-positive nerve terminals exists within the median eminence, in the nucleus paraventricularis, the supraoptic nucleus and the subfornical organ (SFO). Further ANG II-positive brain areas are the substantia gelatinosa of the spinal cord, nucleus tractus spinalis nervi trigemini, nucleus amygdaloideus centralis, sympathetic lateral column, nucleus dorsomedialis hypothalami, locus coeruleus.

The presence of ANG II in the paraventricular nucleus of the hypothalamus (PVN) is of particular interest in view of the capability of ANG II to release adrenocorticotrophic hormone (ACTH) and in view of the projections from the PVN through the external layer of the median eminence to the portal blood circuit.

Lind et al. (38) recently studied the PVN/ANG II system in detail. It was confirmed that antisera to ANG II stain neurosecretory neurons that synthetize vasopressin in magnocellular parts of the PVN, but it was also shown, that ANG II-immunoreactive neurons were scattered throughout the parvocellular division. A subpopulation of ANG II-immunoreactive parvocellular neurons in the PVN projects to the neurohemal zone via the external layer of the median eminence. These ANG II-stained projections were eliminated after bilateral destruction of the PVN. In contrast, the ANG II-stained magnocellular neurons in the PVN appear to project through the internal lamina of the median eminence to the posterior pituitary. Bilateral lesions of the PVN reduce, but do not eliminate ANG II staining in the internal lamina, the remaining fibers probably arising from ANG II-stained magnocellular neurons in the supraoptic nucleus.

These studies suggest that ANG II-stained fibers in the external and internal laminae of the median eminence arise from separate groups of neurons, a conclusion also supported by observations that these areas can be regulated independently. For instance, adrenalectomy leads to a selective enhancement of ANG II-stained fibers in the neurohemal zone as has also been described for CRF, oxytocin, and vasopressin. In contrast, the selective increase of ANG II staining in the internal lamina of the median eminence following water deprivation, supports a functional relationship of this ANG II pathway to the posterior pituitary in vasopressin and oxytocin release.

Another area of the brain, which is of particular interest for the regulation of fluid balance is the subfornical organ (SFO). The SFO is a small glomus-like convexity of the midline third ventricular ependyma near the interventricular foramen. It is densely vascularized, has relatively porous blood brain barrier and is strategically located to monitor plasma, cerebrospinal fluid (CSF) and neuronal inputs. ANG II-receptors have been demonstrated on the SFO (39,40) and stimulation of the receptors in water satiated animals results in copious drinking.

Recently, Lind et al. (37) examined the ANG II pathways to and from the SFO by immunohistochemical methods. ANG II immunoreactive cell bodies and fibers were clearly identified in the SFO of the rat. Cells were distributed in an annulus around the periphery of the SFO. Fibers were observed in a plexus, located centrally within the ring of cells. Knife-cuts through the ventral stalk of the SFO diminished but did not eliminate fiber staining in the SFO. Ventral to the cut, and to a lesser degree, also dorsal to the cut, bright varicose ANG II immunoreactive fibers were described. Combination of immunohistochemistry with retrograde transport, identified the perifornical zone of the lateral hypothalamus, the rostral zona incerta and the nucleus reuniens of the thalamus as the source of ANG II-stained inputs to the SFO, and the region of the median preoptic nucleus as a recipient of ANG II-immunoreactive SFO efferents. It was concluded that ANG II-stained pathways from the lateral hypothalamus and adjacent regions project to the SFO, and that, in turn, ANG II-stained neurons within the SFO project to the preoptic region of the hypothalamus. Interestingly, the perifornical region of the lateral hypothalamic area, and rostral parts of the zona incerta that project to the SFO are known to be involved in the regulation of thirst. Furthermore, the ANG II-stained projection from the SFO to the preoptic region are also thought to play a critical role in the initiation of drinking behavior (40).

The staining was neither altered by water deprivation nor by nephrectomy. Since the former procedure drastically increases and the latter drastically decreases circulating levels of ANG II, tissue bound peptide from the blood does not appear to have been responsible for the observed immunoreactivity. Nevertheless, circulating ANG II may also affect ANG II-receptors in the SFO through the relatively permeable blood brain barrier. Thus, the SFO could function as an integrating center where hormonal and neuronal angiotensin interact to control fluid blance.

Together with previously described data on the localisation of ANG II in the brain a picture begins to emerge which gives further evidence to a role for antiotensin pathways in the brain in blood pressure and volume control.

ACKNOWLEDGEMENTS

This work was supported by the Deutsche Forschungsgemeinschaft (DFG) within the Sonderforschungsbereich "Cardiovasculäres System" (SFB90). The secretarial help of B. Hess is gratefully acknowledged.

REFERENCES

1. Schelling, P., Meyer, D., Loos, H.E., Speck, G., Philipps, M.I., Johnson, A.K., Ganten, D. Neuropharmacology R1: 455-463, 1982.
2. Ganten, D., Hermann, K., Bayer, C., Unger, Th., Lang, R.E. Science 221, 4613: 869-871, 1983.
3. Eggena, P., Ito, T., Barrett, J.D., Villareal, H., Sambhi, M.P. In: Exp. Brain Res. (Suppl 4). (Eds. D. Ganten, M. Printz, M.I. Phillips, B.A. Schölkens), Springer, Berlin, Heidelberg, New York, pp. 169-177, 1982.
4. Ganten, D., Marquez-Julio, A., Granger, P., Hayduck, K., Karsunky, K.P., Boucher, R., Genest, J. Am. J. Physiol. 221: 1733-1737, 1971.
5. McDonald, W., Wickre, C., Aumann, S., Ban, D., Moffit, B. Endocrinology 107: 1305-1308, 1980.
6. Rettig, R., Lang, R.E., Rascher, W., Unger, Th., Ganten, D. Clin. Sci. 63: 269-283, 1982.
7. Unger, Th., Kaufmann-Bühler, I., Schölkens, B., Ganten, D. Eur. J. Pharmacol. 70: 467-478, 1981.
8. Unger, Th., Ganten, D., Lang, R.E. Clin. exp. Hypertens. A5 (7/8):

1333-1354, 1983.

9. Mann, J.F.E., Phillips, M.I., Dietz, R., Haebara, H., Ganten, D. Am. J. Physiol. 234 (5): 629-637, 1978.

10. Evered, M.D., Robinson, M.M., Richardson, M.A. Eur. J. Pharmacol. 68: 443-449, 1980.

11. Berecek, K.H., Okuno, T., Nagahama, S., Oparil, S. Hypertension 5: 689-700, 1983.

12. Reid, I.A., Morris, B.J., Ganong, W.F. Ann. Rev. Physiol. 40: 337-410, 1978.

13. Ganten, D., Printz, M., Phillips, M.I., Schölkens, B.A. (Eds.) In: Exp. Brain Res. (Suppl. 4), Springer, Berlin, Heidelberg, New York, 1982

14. Printz, M., Ganten, D., Unger, Th., Phillips, M.I. In: Exp. Brain Res. (Suppl. 4). (Eds. D. Ganten, M. Printz, M.I. Phillips, B.A. Schölkens) Springer, Berlin, Heidelberg, New York, pp. 2-52, 1982.

15. Misono, K.S., Chang, J.-J., Inagami, T. Clin. exp. Hypertens. A5 (7/8): 941-959, 1983.

16. Rougeon, F., Chambraud, B., Foote, S., Pathier, J.-J., Nageotte, R., Corvol, P. Proc. natn. Acad. Sci. USA 78: 6367-6371, 1981.

17. Imai, T., Miyazaki, H., Hirose, S., Murakami, K. Clin. exp Hypertens. A5 (7/8): 961-967, 1983.

18. Hirose, S., Ohsawa, T., Inagami, T., Murakami, K. J. biol. Chem. 257: 6316-6321, 1982.

19. Hirose, S., Yokosawa, H., Inagami, T., Workman, K.J. Brain. Res. 191: 489-499, 1980.

20. Ganten, D., Speck, G. Biochem. Pharmac. 27: 2378-2389, 1978.

21. Ludwig, G., Lehmann, E., Murakami, K., Lang, R.E., Unger, Th., Ganten, D. Demonstration of specific renin messenger RNA in mouse brain. Submitted for publication. 1984.

22. Nakanishi, S., Ohkubo, H., Nawa, H., Kitamura, N., Kageyama, R., Ujihara, M. Clin. exp. Hypertens. A5 (7/8): 997-1003, 1983.

23. Hilgenfeldt, U. Clin. exp. Hypertens. A5 (7/8): 1021-1035, 1983.

24. Campbell, D.J., Bouhnik, J., Menard, J., Corval, P. Nature 308 (5955): 206-208, 1984.

25. Fuxe, K., Ganten, D., Köhler, C., Schüll, B., Speck, G. Acta. physiol. scand. 110: 321-323, 1980.

26. Strittmatter, S.M., Lo, M.M.S., Javitch, J.A., Snyder, S.H. Proc.

natn. Acad. Sci. USA: in press, 1984.

27. Yang, H.Y.T., Neff, N.H.   J. Neurochem. 19: 2443-2450, 1972.

28. Rix, E., Ganten, D., Schüll, B., Unger, Th., Taugner, R.   Neurosci. Lett. 22: 125-130, 1981.

29. Paul, M., Hermann, K., Printz, M., Lang, R.E., Unger, Th., Ganten, D. J. Hypertens 1 (suppl. 1): 9-15, 1977.

30. Arregui, A., Bennet, J.P., Bird, E.O., Aymamura, H.J., Iversen, L.L., Synder, S.H.   Ann. Neurol. 2: 294-298, 1977.

31. Unger, Th., Ganten, D., Lang, R.E., Schölkens, B.A.   J. Cardiovasc. Pharmacol. 6: 872-880, 1984.

32. Balz, W., Herrmann, K., Unger, Th., Lang, R.E., Ganten, D.   Angiotensin in monkey tissue. In preparation, 1984.

33. Ganten, D., Fuxe, K., Phillips, M.I., Mann, J.F.E., Ganten, U.   In: Frontiers in neuroendocrinology. (Eds. W.F. Ganong, L.D. Marin) Raven Press, New York, pp. 61-99, 1978.

34. Brownfield, M.S., Reid, I.A., Ganten, D., Ganong, W.F.   Neuroscience 7 (7): 1759-1769, 1982.

35. Zimmerman, E.A., Krupp, L., Hoffmann, D., Matthew, E., Nilaver, G. Peptides 1 (Suppl. 1): 3-10, 1980.

36. Fuxe, K., Ganten, D., Hökfelt, T., Bolme, P.   Neurosci. Lett. 2: 229-234, 1976.

37. Lind, R.W., Swanson, L.W., Ganten, D.   Angiotensin II immunoreactivity in the neural afferents and efferents of the subfornical organ of the rat.   Brain Res.: in press, 1984.

38. Lind, R.W., Swanson, L.W., Bruhn, T.O., Ganten, D.   The distribution of angiotensin II immunoreactive cells and fibers in the para-ventriculo-hypophysical system of the rat.   Brain Res.: in press, 1984.

39. Felix, D., Schelling, P., Haas, H.L.   In: Brain Res. (Suppl. 4) The renin angiotensin system in the brain. (Eds. D. Ganten, M. Printz, M.I. Phillips, B.A. Schölkens), Springer, Berlin, Heidelberg, New York, pp. 255-269, 1982.

40. Lind, R.W., Johnson, A.K.   In: Exp. Brain Res. (Suppl. 4). The renin angiotensin system in the brain. (Eds. D. Ganten, M. Printz, M.I. Phillips, B.A. Schölkens), Springer, Berlin, Heidelberg, New York, pp. 353-364, 1982.

8

ENDOGENOUS OPIOIDS AND CENTRAL CARDIOVASCULAR CONTROL

FADEN, A.I., MCINTOSH, T.K.

Department of Neurology, University of California at San Francisco, and Neurology Service, San Francisco Veterans Administration Medical Center, 4150 Clement Street, San Francisco, California 94121, USA

ABSTRACT

A number of endogenous opioid peptides have been identified since the discovery of the pentapeptide enkephalins in 1975. Like exogenous opiates, these substances have potent cardiovascular effects following administration into the central nervous system (CNS). Combined with the observation that endogenous opioids are concentrated in central cardioregulatory sites, such findings have suggested a role for these substances in central cardiovascular regulation. Findings from work in experimental shock and hypertension lend further support for this concept. This review summarizes the recent experimental evidence to support the conclusion that endogenous opioid systems may be important in central cardiovascular control.

INTRODUCTION

A number of endogenous opioid peptides have been identified within the central nervous system (CNS) of mammals since the discovery of the pentapeptide enkephalins in 1975: these substances appear to have diverse physiological roles and may serve as hormones, neurotransmitters, and neuromodulators (1-4). There is evidence to support a potential role for endogenous opioids in analgesia, immune control, feeding behavior, respiration, thermoregulation, gastro-intestinal motility, and cardiovascular function (2). Like exogenous opiates, endogenous opioids have potent cardiovascular effects following administration either systemically or into the CNS (5). Moreover, endogenous opioids have been implicated in the pathophysiology of shock and possibly hypertension (2, 6). The present review summarizes the experimental evidence to support a role for endogenous opioid peptides in central cardiovascular regulation.

ENDOGENOUS OPIOIDS AND OPIATE RECEPTORS

The majority of endogenous opioid peptides fall into three classes, each derived from a distinct prohormone precursor: (1) preproenkephalin A, from which methionine-enkephalin, leucine-enkephalin, and larger enkephalin fragments are derived; (2) preproenkephalin B (preprodynorphin), from which dynorphins, neo-endorphins, and leucine-enkephalin are derived; and (3) preproopiomelanocortin, from which ß-endorphin is derived (7). These systems have distinct distributions within the CNS, although they overlap in some areas. Less information is available regarding the distribution and derivation of other opioid peptides such as kyotorphin, demorphin, and FMRFamide.

Six classes of opiate receptors have been proposed (μ, δ, κ, ϵ, σ, and λ), with strong evidence to support the existence of at least μ, δ and κ receptors (2). Such receptors have usually been defined on the basis of bioassays or cross-tolerance studies. There is also some experimental support for the existence of isoreceptors, for several of these receptor classes, particularly for κ and μ (8-10). The complexity of the endogenous opioid system is evident from the fact that there is no clear relationship between the three endogenous opioid systems and specific opiate receptors. Thus, ß-endorphin shows affinity for μ, δ and κ receptors, as well as for the putative ϵ receptor. Enkephalins act at both μ and δ receptors. Dynorphins act at both μ and κ receptors, with the shorter fragments having increased activity at μ-sites and the larger fragments at κ-sites (7).

Endogenous opioids and opiate receptors have been found in a large number of central cardioregulatory sites including limbic cortex, hypothalamus, brainstem, and spinal cord (11-19). The physiological roles of endogenous opioids, including their potential role in cardiovascular function, are based on three kinds of studies: (1) examination of the effects of local injections of endogenous opioids within the CNS; (2) examination of the effects of opiate receptor antagonists to infer the action of endogenous opioids; and (3) correlation of changes in endogenous opioid immunoreactivity or changes in opiate receptor regulation with pathophysiological conditions such as shock or hypertension.

EFFECTS OF CENTRALLY INJECTED ENDOGENOUS OPIOIDS ON CARDIOVASCULAR FUNCTION

It has been known for many years that exogenous opiates like morphine produce cardiovascular effects following injection into the CNS. Following discovery of the endogenous opioids, many groups have examined cardiovascular and respiratory effects of centrally administered opioid peptides and analogs. Some of this data has appeared to be conflicting, with considerable confusion regarding specific roles (e.g., hypertensive or hypotensive) of the opioids. Much of the confusion and conflict relates to methodological issues that complicate interpretation of these studies. Such critical issues include choice of species, site of administration, anesthetic state, type of anesthetic, respiratory state, type of opioid ligand, and choice of dose.

Several groups have observed that opioid peptides or analogs may have differing effects in various species. Thus, Laubie et al. found that the stable synthetic agonist (D-Ala^2)Met-enkephalinamide (DAME) produced hypotension and bradycardia following intracisternal injection in anesthetized dogs (20); in contrast, Bellet et al. found that DAME caused an increase in blood pressure and heart rate following intracisternal injections in anesthetized rats (21). Similarly, Hassen and co-workers found that the synthetic agonist (D-Ala^2, D-Leu^5)-enkephalin (DADL) elicited dose-dependent decreases in blood pressure and heart rate following injections into the nucleus tractus solitarius (NTS) in anesthetized cats (22), but produced dose-dependent increases in blood pressure and heart rate following NTS injections in anesthetized rats (23). The opposite effects in the cat and rat each were readily antagonized by the non-specific opiate-receptor antagonist naloxone (22,23). In addition, opposite cardiovascular effects were found for synthetic enkephalins administered into the nucleus ambiguus of ventilated rats and dogs (24,25).

The site of opioid administration also may determine its cardiovascular and/or respiratory effects. Thus, injection of µ- or δ- selective opiates or opioids in rats may produce different cardiorespiratory effects depending upon whether injections are made into the third or fourth cerebral ventricle (26). Similarly, intracisternal and intracerebroventricular (i.c.v.) administration of ß-endorphin in cats may produce different effects on blood pressure and heart rate (27). The importance of site specificity has been further underscored by the observations of Feuerstein and Faden, who demonstrated that μ- or δ-selective opioid agonists caused increases in blood pressure and heart rate following injection into the medial preoptic nucleus of the hypothalamus, whereas the same compounds produced hypotension following microinjection into an adjacent hypothalamic nucleus (periventricularis hypothalami) (28). This latter observation also raises questions about studies utilizing ventricular injections, since microinjections of identical opioids into adjacent periventricular sites, each with access to the third ventricle, produce opposite cardiovascular effects. Therefore, the effects of ventricular injections may represent the summation of a variety of cardiovascular actions from a number of periventricular sites, and thus may not be used to conclude that an opioid has a specific central cardiovascular action.

It has become increasingly clear that the anesthetic state plays a critical role in determining the cardiovascular consequences of centrally administered opiates or opioids. This has been shown following either ventricular or intraparenchymal injections. Thus, Holaday noted different responses in anesthetized and unanesthetized rats after third or fourth ventricular injections of μ and δ agonists (26). Yukimura et al. found that administration of DAME into the lateral cerebral ventricle produced dose-dependent increases in blood pressure in unanesthetized animals, whereas the same dose produced dose-dependent decreases in blood pressure in α-chloralose-anesthetized rats (29). The effects found by Yukimura and colleagues in the unanesthetized rat (29) are also opposite to those found by Wong et al. in pentobarbital-anesthetized rats administered DAME (30). Similarly, Pfeiffer et al. (31, 32) and Faden and Feuerstein (33) noted opposite effects of μ- and δ-selective opioids following microinjections into the anterior hypothalamic region in unanesthetized versus pento-barbital-anesthetized rats. Whether differing cardiovascular effects of CNS opioid administration result from choice of various anesthetics has not been well studied. However, several anesthetics have been found to cause naloxone-reversible changes (34, 35), an observation that may confound interpretation of cardiovascular changes following opioid administration in anesthetized animals.

Florez and Mediavilla first demonstrated that enkephalins may produce respiratory depression following administration onto the ventral surface of the brainstem in anesthetized cats (36). Subsequently, it has been recognized that the respiratory state may critically determine the cardiovascular effects of centrally administered opioids. Thus, Bellet et al. found that high doses of morphine and DAME caused a biphasic effect with secondary hypotension and bradycardia in spontaneously breathing rats, but that DAME produced only a pressor response in artificially ventilated rats (21). Hassen et al., in several independent studies, have also demonstrated that various opioid agonists produce opposite effects in spontaneously breathing and artificially respired animals (8,24,37).

The type of opioid utilized in any particular study is also of considerable importance in determining cardiovascular or respiratory effects of centrally administered opioids. Thus, Holaday found opposite effects of μ- and δ-selective ligands following administration to third and fourth cerebral ventricles (26). Bolme et al. found that ß-endorphin produced preferential vasodepressor responses and bradycardia, whereas methionine- and leucine-enkephalin produced vasopressor responses and bradycardia after intracisternal administration in anesthetized rats (38). McDonald et al. similarly observed that i.c.v. injection of the μ-agonist morphiceptin produced hypotension and bradycardia, whereas i.c.v. injection of the δ-agonist DADL produced tachycardia (39). Pfeiffer et al. observed different cardiovascular responses produced by μ-, δ- and κ-selective opiate agonists following microinjection into the anterior hypothalamic region of unanesthetized rats (31). Faden and Feuerstein observed that specific opiate-receptor agonists may produce different effects following microinjection into the hypothalamic region of anesthetized rats (33). Petty and De Jong observed that ß-endorphin produced profound hypotension, whereas DAME increased blood pressure following microinjections in the NTS of the anesthetized rat (40). Petty and Reid observed that μ-selective opiate agonist reduced baroreceptor sensitivity, whereas κ- and δ-opiate agonists increased baroreceptor sensitivity in the pentobarbital-anesthetized rabbit (41). Hassen et al. found that μ- and κ-selective opiate agonists produced differing responses following microinjection into hindbrain nuclei of anesthetized rats (37). Interestingly, different opioids having the same opiate selectivity have occasionally been found to produce differing responses, as shown by Hassen et al. (8) and by Faden and Feuerstein (33). These differences may have resulted from differing potencies and thus choice of doses (vide infra); however, these observations may also suggest that various agonists act on different isoreceptors (9,10) that produce variable cardiovascular actions.

It has long been recognized that the dose of an exogenous opiate may determine whether that opiate produces hypertension and tachycardia on the one hand, or hypotension and bradycardia on the other hand. This has also been observed with regard to administration of endogenous opioids or synthetic opioid peptides. Thus, Bellet et al. noted that low doses of DAME caused a pressor response with tachycardia, whereas higher doses produced a biphasic effect with secondary hypotension and bradycardia (21). Pfeiffer et al. have shown that μ- and δ-selective agonists produce an inverted U-shaped dose-response with regard to blood pressure and heart rate actions (31). The importance of dose has also been stressed by Faden and Feuerstein with regard to μ- and δ-selective enkephalins following hypothalmic microinjection (33), by Petty and De Jong with regard to ß-endorphin following injection into the NTS (40), and by Hassen et al. with regard to μ- and δ-selective enkephalins following injection into the NTS (23). Thus, failure to examine complete dose-response curves may in certain studies have resulted in conflicting conclusions about the cardiovascular effects of specific opioids. For example, Bolme et al. observed that DAME produced hypotension and bradycardia following intracisternal administration in rats (38). However, the dose they used was relatively high; subsequently, Bellet and colleagues (21) demonstrated a biphasic response utilizing a more complete dose-response study, finding that DAME produced preferentially vasopressor responses at much lower doses than utilized by Bolme et al. (38). Therefore, adequate evaluation of the cardiovascular actions of a specific opioid in a specific site may require examination of doses varying over five or six orders of magnitude.

CARDIOVASCULAR EFFECTS OF OPIOIDS AS SUGGESTED FROM STUDIES OF OPIATE ANTAGONISTS OR STUDIES OF OPIATE RECEPTOR REGULATION

The use of either non-selective or selective opiate and receptor antagonists to infer the existence of endogenous opioids has been best utilized in studies of shock (vide infra). In general, the cardiovascular effects produced by μ-selective agonists are most readily antagonized by naloxone, although naloxone at higher doses may reverse the cardiovascular effects of all classes of opioid agonists. Other receptor-selective antagonists such as the μ-selective antagonist naloxazone (9), the μ-selective antagonist funaltrexamine (32), or the κ-selective antagonist MR2266 (32) have been utilized to provide support for conclusions regarding the cardiovascular actions of specific opiate receptor populations. However, injections of opiate antagonists in normal animals, which have not been subjected to stressors or pathophysiological stimuli, produce minimal effects, suggesting that opioid systems may not be tonically active but rather potentially active or reactive (42). Nonetheless, the importance of opiate receptors in mediating the action of opioids has been well demonstrated by Pfeiffer et al. who noted that up-regulation of opiate receptors following chronic naloxone administration enhanced the cardiovascular depressant effects but not the respiratory depressant effects of opiates (43).

ENDOGENOUS OPIOIDS AND OPIATE RECEPTORS IN SHOCK AND HYPERTENSION

Holaday and Faden first proposed a role for endogenous opioids in the pathophysiology of shock, demonstrating that the opiate antagonist naloxone improved cardiovascular function and outcome following shock resulting from endotoxemia (42,44), hypovolemia (45) or spinal cord damage (46,47) in a variety of species. They further demonstrated that the beneficial cardiovascular effects of naloxone in these models were stereospecific (44,46), supporting the conclusion that the effects of the opiate antagonist were mediated by opiate receptors. These authors also found that lateral ventricular injections of naloxone, at doses that were ineffective when given systemically, reserved cardiovascular depression produced by spinal transection, indicating that the effects of naloxone in this model were centrally mediated (47). Utilizing ventricular-cisternal perfusion, Janssen and Lutherer also demonstrated that the beneficial cardiovascular effects of naloxone endotoxin shock in dogs were mediated centrally (48). Moreover, Amir found that the protective effects of the opiate antagonist naltrexone in anaphylactic shock were not duplicated by a quaternary derivative of naltrexone that does not cross the blood-brain barrier (49). Taken together, these studies indicate that cardiovascular depression in shock is caused, in part, by endogenous opioids acting on the central nervous system.

Further confirmation for this hypothesis comes from the observations that there are changes in levels of endogenous opioids in brain parenchyma, as well as changes in opiate receptors, following shock (50,51). Use of receptor-selective antagonists and dose-response studies of naloxone have suggested that the beneficial effects of opiate antagonists in shock are mediated by non- μ opiate receptors (6,52).

A potential role for endogenous opioids in the pathophysiology of hypertension has been inferred from several lines of evidence. Zamir et al. noted that experimentally hypertensive rats appeared to be less sensitive to nociceptive stimuli than control animals; this increased antinociception was reversed by naloxone (53). Similarly, Wendel and Bennett found that spontaneously hypertensive rats showed less nociceptive sensitivity than control Wister-Kyoto (WKY) rats, even in young animals prior to the establishment of hypertension (54). In addition, several groups have found that SHR rats have increased sensitivity to the cardiovascular effects of centrally administered opioids. Thus, Yukimura and colleagues noted that SHR rats showed an increased pressor response to centrally administered DAME than WKY rats (29). Similarly, Schaz et al. found that the blood pressure increases following intraventricular administration of leucine-enkephalin was significantly greater in SHR rats than normotensive controls (55). In addition, Feuerstein et al. found that SHR rats were more sensitive than WKY rats to the cardiovascular effects of (D-Ala^2,MePhe^4-Gly^{5}ol)-enkephalin, a μ-selective agonist, following microinjection into the hypothalamus (56). Moreover, concentrations of dynorphin and leucine-enkaphalin differed between SHR and WKY rats in various brain nuclei and the pituitary gland.

The precise mechanisms by which endogenous opioids produce their central cardiovascular effects remain to be fully elucidated. Various studies have pointed to interactions with the sympathetic nervous system (32,57-60), parasympathetic nervous system (25,32,46,61), or baroreceptor reflex (41,55), depending upon type of opioid, site of administration and dose.

CONCLUSIONS

Considerable recent evidence indicates that endogenous opioids may play a role in central cardiovascular regulation, both with regard to homeostatic control and under pathophysiological conditions. Endogenous opioids and opiate receptors are found in central cardioregulatory sites, and endogenous opioids produce cardiovascular changes following extremely small doses administered into the central nervous system. Cardiovascular effect of endogenous opioids are largely reversed by opiate antagonists, thus suggesting an opiate receptor mechanism of action. The specific cardiovascular actions observed following central administration depend upon choice of species, site of injection, anesthetic state, respiratory state, type of opioid administered, and the dose of the opioid. Given these variables, it is difficult to draw non-contingent conclusions about the specific roles of endogenous opioids in central cardiovascular regulation. In unanesthetized animals, μ- or δ-selective agonists in the forebrain generally produce hypertension and tachycardia (31,55). In anesthetized animals, these effects appear to be altered in favor of a cardiodepressant action, including hypotension and bradycardia (26,28,33). The effects of μ-and δ-selective agonists following hindbrain injections are less consistent, being highly dependent upon site of administration, and probably anesthetic state (22-24,28,62). κ-selective agonists, particularly dynorphin A-(1-13) and dynorphin A-(1-17), appear to have largely cardiodepressant actions in both forebrain and hindbrain (8,63). These effects of dynorphin, combined with the high doses of naloxone required to treat experimental shock, suggest that this opioid may be a pathophysiological factor in shock. Studies of experimental shock and experimental hypertension further suggest a role for endogenous opioids in the pathophysiology of these conditions. However, the specific roles of individual opioids and the precise mechanisms of action have yet to be completely elucidated.

ACKNOWLEDGEMENT

We whish to thank Miss Eleanor M. Bell for prepartion of this manuscript.

REFERENCES

1. Bloom, F.E. Annu. Rev. Pharmacol. Toxicol. 23: 151-170, 1985.
2. Faden, A.I. JAOA 84 (Suppl 1): 129-134, 1984.
3. Frederickson, R.C., Geary, L.E. Prog. Neurobiol. 19: 19-69, 1982.
4. Henry, J.L. Neurosci. Biobehav. Rev. 6: 229-245, 1982.

5. Holaday, J.W.   Annu. Rev. Pharmacol. Toxicol. 23: 541-594, 1983.

6. Faden, A.I.   JAMA 252: 1177-1180, 1984.

7. Cox, B.M.   Life Sci. 31: 1645-1658, 1982.

8. Hassen, A.H, Feuerstein, G., Faden, A.I.   J. Neurosci. 4: 2213-2221, 1984.

9. Holaday, J.W., Pasternak, G.W., Faden, A.I.   Neurosci. Lett. 37: 199-204, 1983.

10. Wuster, M., Schulz, R., Herz, A.   Biochem. Pharmacol. 30: 1983, 1981.

11. Della Bella, D., Casacci, F., Sassi, A.   Adv. Biochem. Phsycho-pharmacol. 18: 271-277, 1978.

12. Dupont, A., Lepine, J., Langelier, P., Merand, Y., Rouleau, D., Vaudry, H., Gros, C., Barden, N.   Reg. Peptides 1: 43-52, 1980.

13. Elde, R., Hokfelt, T., Johansson, O., Terenius, L.   Neuroscience 1: 349-351, 1976.

14. Feuerstein, G., Molineaux, C.J., Rosenberger, J.G., Faden, A.I, Cox, B.M.   Peptides 4: 225-229, 1983.

15. Hokfelt, T., Elde, R., Johansson, O., Terenius, L., Stein, L. Neurosci. Lett. 5: 25-31, 1977.

16. Hughes, J., Kosterlitz, H.W., Smith, T.W.   Br. J. Pharmacol. 61: 639-648, 1977.

17. Khachaturian, H., Watson, S.J., Lewis, M.E., Coy, D., Goldstein, A. Akil, H.   Peptides 3: 941-945, 1982.

18. Pfeiffer, A., Herz, A.   Molec. Pharmacol 21: 266-271, 1982.

19. Simantov, R., Kuhar, M.J., Uhl, G.R., Snyder, S.H.   Proc. natn. Acad. Sci. USA 74: 2167-2172, 1977.

20. Laubie, M., Schmitt, H., Vincent, M., Remond, G.   Eur. J. Pharmacol. 46: 67-71, 1977.

21. Bellet, M., Elghozi, J.L., Meyer, P., Pernollet, M.G., Schmitt, H. Br. J. Pharmacol. 71: 365-369, 1980.

22. Hassen, A.H., Feuerstein, G., Pfeiffer, A., Faden, A.I. Reg. Peptides 4: 299-309, 1982.

23. Hassen, A.H., Feuerstein, G., Faden, A.I.   Peptides 3: 1031-1037, 1982.

24. Hassen, A.H., Feuerstein, G., Faden, A.I.   Neuropharmacology 23: 407-415, 1984.

25. Laubie, M., Schmitt, H., Vincent, M.   Eur. J. Pharmacol. 59:

287-291, 1979.

26. Holaday, J.W. Peptides 3: 1023-1029, 1982.
27. Feldberg, W., Wei, E. J. Physiol. 275: 18P, 1978.
28. Feuerstein, G., Faden, A.I. Life Sci. 31: 2197-2200, 1982.
29. Yukimura, T., Unger, T., Rascher, W., Lang, R.E., Ganten, D. Clin. Sci. 61: 347s-350s, 1981.
30. Wong, T.M., Chan, S.H.H., Tse, S.Y.H. Neurosci. Lett. 46: 249-254, 1984.
31. Pfeiffer, A., Feuerstein, G., Kopin, I.J., Faden, A.I. J. Pharmac. exp. Ther. 225: 735-741, 1983.
32. Pfeiffer, A., Feuerstein, G., Zerbe, R.L., Faden, A.I., Kopin, I.J. Endocrinology 113: 929-938, 1983.
33. Faden, A.I., Feuerstein, G. Br. J. Pharmacol. 79: 997-1002, 1983.
34. Arndt, J.O., Freye, E. Nature 277: 399-400, 1979.
35. Berkowitz, B.A., Ngai, S.H., Finch, A.D. Science 194: 967-968, 1976.
36. Florez, J., Mediavilla, A. Brain Res. 138: 585-590, 1977.
37. Hassen, A.H., Feuerstein, G., Faden, A.I. Peptides 4: 621-625, 1983.
38. Bolme, P., Fuxe, K., Agnati, L.F., Bradley, R., Smythies, J. Eur. J. Pharmacol. 48: 319-324, 1978.
39. McDonald, W.J., Motter, M., Ganten, D., Lang, R.E. Clin. exp. Hypertens. 10: 1837-1841, 1984.
40. Petty, M.A., De Jong, W. Eur. J. Pharmacol. 81: 449-457, 1982.
41. Petty, M.A., Reid, J.L. Naunyn-Schmiedebergs exp. Path. Arch. Pharmak. 319: 206-211, 1982.
42. Holaday, J.W., Faden, A.I. Nature 275: 450-451, 1978.
43. Pfeiffer, A., Pfeiffer, D.G., Feuerstein, G., Faden, A.I., Kopin, I.J. Brain Res. 296: 305-311, 1984.
44. Faden, A.I., Holaday, J.W. J. Pharmac. exp. Ther. 212: 441-447, 1980.
45. Faden, A.I., Holaday, J.W. Science 205: 317-318, 1979.
46. Faden, A.I., Jacobs, T.P., Holaday, J.W. J. Auton. Nerv. Syst. 2: 295-304, 1980.
47. Holaday, J.W., Faden, A.I. Brain Res. 189: 295-299, 1980.
48. Janssen, H.F., Lutherer, L.O. Brain Res. 194: 608-612, 1980.
49. Amir, S. Eur. J. Pharmacol. 80: 161-162, 1982.
50. Feuerstein, G., Faden, A.I., Krumins, S.A. Eur. J. Pharmacol. 100: 245-246, 1984.

51. Feuerstein, G., Molineaux, C.J., Rosenberger, J.G., Zerbe, R.L., Cox, B.M., Faden, A.I. Am. J. Physiol. 249: E244-E250, 1985.

52. Feuerstein, G., Powell, E., Faden, A.I. Peptides 6 (Suppl 1): 11-13, 1985.

53. Zamir, N., Simantov, R., Segal, M. Brain Res. 184: 299-310, 1980.

54. Wendel, O.T., Bennett, B. Life Sci. 29: 515-521, 1981.

55. Schaz, K., Stock, G., Simon, W., Schlor, K.H., Unger, T., Rockhold, R. Ganten, D. Hypertension 2: 395-407, 1980.

56. Feuerstein, G., Zerbe, R.L., Faden, A.I. Hypertension 5: 663-671, 1983.

57. Farsang, C., Ramirez-Gonzalez, M.D., Mucci, L., Kunos, G. J. Pharmac. exp. Ther. 214: 203-208, 1980.

58. Holaday, J.W., D'Amato, R.J., Ruvio, B.A., Feuerstein, G., Faden, A.I. Circ. Shock 11: 201-210, 1983.

59. Van Loon, G.R., Appel, N.M., Ho, D. Brain Res. 212: 207-214, 1981.

60. Simon, W., Schaz, K., Ganten, U., Stock, G., Schlor, K.H., Ganten, D. Clin. Sci. mol. Med. 55: 237s-241s, 1978.

61. Laubie, M., Schmitt, H. Eur. J. Pharmacol. 71: 401-409, 1981.

62. Freye, E., Hartung, E., Schenk, G.K. Pharmacology 25: 6-11, 1982.

63. Feuerstein, G., Faden, A.I. Neuropeptides 5: 295-298, 1984.

PART THREE

CLINICAL ASPECTS

9

EFFECTS OF VOLUNTARY CONTROL OF BLOOD PRESSURE ON CARDIOVASCULAR REGULATION DURING POSTURAL CHANGE

SUTER, T.W., WEIPERT, D., SHAPIRO, D.

Department of Psychiatry, University of California, 760 Westwood Plaza, Los Angeles, CA 90024

ABSTRACT

This research is part of a larger program on the behavioral and psychological regulation of cardiovascular adjustments to postural change. The long-range objective is to develop methods for the behavioral management of patients with orthostatic hypotension. This paper describes methods developed for the continuous non-invasive tracking of systolic and diastolic blood pressure as well as heart rate during postural change --- going from a sitting to a standing position, and the use of these methods in two studies of young, healthy, normotensive individuals. This first study shows the methods to be sensitive to blood pressure and heart rate differences in postural regulation as a function of family history of hypertension. The second shows that cardiovascular adjustments to postural change can be modified by voluntary control and biofeedback procedures. The research supports the role of the central nervous system in the regulation of basic circulatory reflexes.

This research is part of a larger program on the behavioral and psychological regulation of cardiovascular adjustments to postural change. The long-range objective is to develop methods for the behavioral management of patients with crthostatic hypotension. In our initial studies, we developed non-invasive methods for assessing blood pressure and heart rate changes during postural adjustments in healthy young male volunteers. Two sets of findings will be summarized. The first concerns the relation between family history of hypertension and orthostatic stress response, and the second concerns the ability of subjects to alter the pattern of cardiovascular change during orthostatic stress by voluntary control and biofeedback methods.

Assumption of upright posture causes blood pooling in the extremities and a decrease in cardiac output (1). The fall in blood pressure is buffered by a compensatory increase in sympathetic nervous system activity (2) which produces vasoconstriction and tachycardia, secretion of adrenomedullary catecholamines (3), and activation of the renin-angiotensin system (4). These responses are triggered by reflexes arising in the arterial baroreceptors, cardiopulmonary, vestibular, and somatic afferents (5). Reflex control of circulation plays an important role in maintaining arterial blood pressure during postural stress.

The baroreflex (BR) is considered to be the most potent neural regulator of blood pressure, acting to stabilize phasic changes in a homeostatic fashion. The arterial baroreceptors are nerve endings in the adventitia of the walls of large blood vessels (6). The aortic BR acts predominantly as an anti-hypertensive mechanism, whereas the carotid BR is more effective in buffering a decrease in pressure below normal level (7). The onset of this mechanism is immediate upon a change in blood pressure, normally within one or two heart beats. A BR latency period of about 0.6 seconds has been reported (8).

Sensitivity of the BR (BRS) is expressed as rate of change of heart period as a function of change in systolic blood pressure (msec/mmHg). Studies of BRS have been performed under different experimental conditions: bicycling (9) and mental arithmetic (10,11) in normotensive subjects, and operant conditioning of blood pressure in baboons (12) and heart rate in monkeys (13). The results of these studies indicate a diminution of BRS with increased muscular and/or neural activity.

A non-invasive tracking system was used to record blood pressure (14); systolic blood pressure (SBP) was recorded on the left upper arm; diastolic blood pressure (DBP) was recorded on the right upper arm. Korotkoff (K)-sounds were recorded with piezoelectric microphones, each taped on the skin over the brachial artery. With the cuff pressure set at about the subject's initial SBP, if a K-sound occurs, the subject's SBP is higher than the pressure in the cuff, and the cuff pressure is increased by 3 mmHg. Conversely, if a K-sound is absent, the cuff pressure is decreased by the same amount. The cuff pressure changes are made rapidly at 375 msec following each successive R-wave in the EKG. The same method was used for tracking DBP except that the cuff pressure was decreased by 3 mmHg when a K-sound occurred, and increased by the same amount when a K-sound was absent. In general, the tracking cuff system provides blood pressure data that correspond closely with those obtained by intra-arterial recording (14). Heart rate (HR) and respiration rate (RR) were also recorded continuously.

Subjects were seated in a straight-back chair with their arms extended downward in the same way both during sitting and standing conditions. They simply stood up when instructed to by a computer display, and were asked to avoid excessive movements and thereby minimize artifacts.

In our first study our objectives were: (1) to assess beat-by-beat changes in SBP, DBP, HR, and RR during postural change (sitting to standing) in healthy young subjects; (2) to compare level, amplitude and timing of the orthostatic stress response in indivuals with and without a family history of hypertension; (3) to examine BRS during orthostatic stress. The data were obtained in three successive trials of going from sitting to standing, in 60 subjects, 30 with and 30 without a family history of hypertension (high and low risk, respectively).

Orthostatic stress has been used in comparative studies of cardiovascular reactivity in hypertensives and normotensives, but its role as a predictor of hypertension is not established (15). With provocative mental stressors and physical exercise, greater cardiovascular reactivity, possibly genetically determined, has been shown in indivuals with a family history of hypertension, as compared with indivuals without this risk factor.

The average responses of the four physiological variables during the three trials for low and high risk subjects are shown in Figure 1. The curves reflect the characteristic phasic response of SBP and DBP with an initial fall and a subsequent recovery and a mirror-image response in HR and RR. The BP response patterns conform quite well to those obtained with arterial catheterization (16). In general, the overall reponse patterns were similar in the two hypertension-risk groups. HR tended to be higher during the standing phase and the peak HR occurred later in low risk subjects. SBP was lower and DBP tended to be higher for low risk subjects in this phase. There were no group differences for mean arterial pressure, but analysis of the pulse pressure (SBP-DBP) data indicated that high risk subjects showed less of a drop and started recovery earlier than low risk subjects. Analysis of BRS, however, did not reveal significant group differences. The average values were 11.7 + 3.8 msec/mmHg for the high risk group and 10.3 + 6.0 msec/mmHg for the low risk group.

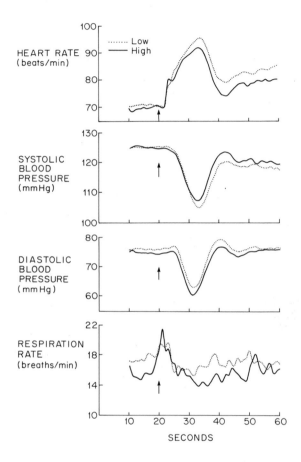

Figure 1. Second-by-second physiological responses averaged over three trials of postural change (sitting to standing) for thirty high risks (positive family history of hypertension) and thirty low risks (negative family history) subjects. At the arrow, the signal was given to stand up.

Thus, we have evidence that the two groups exhibited different adaptation levels to standing. Low risk of hypertension subjects showed higher heart rate and lower systolic pressure during the recovery phase, indicating a better match with normative data for the later phase of postural regulation (16,17). This phase is associated with decreased pulse pressure at the same level of mean arterial pressure and an increased pulse rate of about +15 bpm above pre-standing level.

The pattern of results suggests that the later phase is characterized by decreased variability or adaptability of peripheral resistance in subjects with a family history of hypertension as evidenced by a larger pulse pressure and a lower heart rate. In examining the predictive power of the response to orthostatic stress in young healthy indivuals, greater attention should be focused on the late phase of adaptation to the standing posture (static adaptation), and physiological recordings should be made over a longer period of orthostatic stress. In an older population (over 40 years), the initial phase (dynamic adaptation) may be suitable inasmuch as differences in baroreflex sensitivity are more likely in this age group (21).

In the second study, 40 healthy male college student volunteers were asked to increase their diastolic pressure or their heart rate while sitting and during orthostatic stress (going from a sitting to a standing position), and half of them were also given sec-to-sec visual feedback for the target variable (22). Systolic blood pressure was also continuously recorded. Comparisons were made between baseline and voluntary control conditions, and test trials were included to examine immediate carry-over effects. Substantial increases in tonic levels were obtained for the three cardiovascular variables in both sitting and postural change conditions. The increases were significantly greater for feedback than for no-feedback conditions. Phasic effects of feedback were also observed. The blood pressure troughs and the heart rate peak occurred earlier with feedback. Immediate transfer effects in voluntary control test trials were obtained only in feedback conditions, In general, the results indicate that substantial voluntary control is possible, both of increases in tonic levels as well as of the timing of phasic changes in SBP, DBP, and HR. These effects are facilitated when subjects are fed back information about their own responses.

This study lends support to the potential of biofeedback and voluntary control procedures in altering cardiovascular adjustments to orthostatic stress. Such procedures may be useful in the management of patients with postural hypotension with the possibility that different kinds of biofeedback and voluntary control methods may be relevant to different kinds of orthostatic hypotension. The results are in general agreement with other research indicating that the cardiovascular adjustments to exercise and to different stressful stimuli (noxious events, cold pressor, electrical stimulation) can be modified through mediation of the central nervous system. The mechanisms of these modulations need further investigation.

In conclusion, the data on voluntary and learned control of responses to orthostatic stress support the role of the central nervous system in the regulation of basic circulatory reflexes and the potential of the methods for the management of orthostatic hypotension. In addition, the data on the effects of family history of hypertension on responses to orthostatic stress indicate that such factors also play a role in orthostatic regulation.

REFERENCES

1. Sjostrand, T. Acta physiol. scand. 26: 312-327, 1952.

2. Wallin, B.G. Psychophysiology 18: 470-476, 1981.

3. Robertson, D., Johnson, G.A., Robertson, R.N., Nies, A.S., Shand, D.G., Oates, J.A. Cirulation 59: 637-643, 1979.

4. Zanchetti, A.S. Circulation 56: 691-698, 1977.

5. Abboud, F.M., Heistad, D.D., Mark, A., Schmid, P.G. Prog. cardiovasc. Dis. 18: 371-403, 1976.

6. Kirchheim, H.R. Physiol. Rev. 56: 100-176, 1976.

7. Downing, S.E. In: Handbook of Physiology, Section 2: The Cardiovascular System, (Vol. 1, Eds. R.M. Berne, N. Sperelakis & S.R. Geiger) Bethesda, pp. 621-652, 1979.

8. Borst, C. & Karemaker, J. Journal of Autonomic Nervous System 9: 399-403, 1983.

9. Bristow, J.D., Brown, E.B., Cunningham, D.J.C., Howson, M.G., Peterson, E.S., Pickering, T.G. & Sleight, P. Circulation. Res. 28: 582-592, 1971.

10. Brooks, D., Fox, P., Lopez, R. & Sleight, P. Proceedings of the Physiological Society 280: 75P-76P, 1978.

11. Conway, J., Boon, N., Vann Jones, J. & Sleight, P. Hypertension

5: 746-748, 1983.

12. Goldstein, D.S., Harris, A.H. & Brady, J.V. Biofeedback and Self-Regul. 2: 127-138, 1977.

13. Engel, B.T. & Joseph, J.A. Psychophysiology 19: 609-614, 1982.

14. Shapiro, D., Greenstadt, L., Lane, J.D. & Rubinstein, E. Psychophysiology 18: 129-136, 1981.

15. Julius, S. & Schork, M.A. Ann. N. Y. Acad. Sci. 304: 38-52, 1978.

16. DeMarees, H. Cardiology 61: (Suppl. 1): 78-90, 1976.

17. Hull, D.H., Wolthius, R.A., Cortese, T., Longo, M.R. & Triebwasser, J.H. Am. Heart J. 94: 414-420, 1977.

18. Thulesius, D. & Ferner, U. Z. Kreislaufforsch. 61: 742-754, 1972.

19. DeMarees, H. Aerztl. Forsch. 24: 249-256, 1970.

20. Miyamoto, Y., Tamura, T., Nakamura, T., Higuchi, J. & Mikami, T. Jap. J. Physiol. 32: 245-258, 1982.

21. Gribbin, B., Pickering, T.G., Sleight, P. & Peto, R. Circulation Res. 29: 424-431, 1971.

22. Weipert, D., Shapiro, D., Suter, T.W. Psychophysiology, in press.

10

DIFFERING ASPECTS OF PRE-STROKE AND POST-STROKE HYPERTENSION

SCHULTE, B.P.M.(1), LEYTEN, A.C.M.(2), HERMAN, B.(2)

(1) Institute of Neurology, Catholic University, Nijmegen; (2) General Hospitals, Tilburg, The Netherlands

ABSTRACT

Some data on pre-stroke and post-stroke hypertension in a descriptive epidemiological study of stroke in The Netherlands are presented.
All stroke cases in a community of 152.063 inhabitants were registered from 1978 to 1981.
In the 737 strokes occurring during the study period (cerebral infarction 82 %, intracranial hemorrhage 15 %) 59 % of patients had a history of pre-stroke hypertension. Almost no difference in percentages of pre-stroke hypertension was noted among the major types of stroke. Shortly after their stroke 93 % of assessed cases were markedly hypertensive. Post-stroke hypertension was found less frequently when viewing diastolic pressures than when considering the combined systolic and diastolic ones. The overall distributions of combined systolic and diastolic blood pressure did not differ significantly by stroke type. No association was observed between stroke type and level of systolic pressure. Post-stroke hypertension declined during the next days after stroke, gradually approaching a stable baseline. If treated, it proved to be refractory to control by antihypertensive drugs.
The significantly differing percentages of pre- and post-stroke hypertension seem to reflect the direct influence of stroke on cerebral mechanism of blood pressure control.

In 1658 Johann Jakop Wepfer, physician to the Duke of Württemberg, was first to point at brain-heart relationship in stroke. On the basis of clinical signs and autopsies he concluded that those most liable to apoplexy were the obese, those whose face and hands are livid and those whose pulse is constantly unequal, implicating for the first time that hypertensive patients with cardiac disease are prone to stroke (1,2).

Since more than a decade it is widely accepted that hypertension at any age, in either sex, whether systolic or diastolic, and with or without cardiac impairments is the most powerful risk factor for stroke (3). It is demonstrated likewise that treatment of hypertension prevents stroke. This has been shown in at least five large studies (4-8). Thus no serious disagreement exists concerning the causal role of hypertension in stroke or the benefit of treatment of hypertension in the primary prevention of stroke.

In contrast with pre-stroke hypertension, the role of immediate (i.e. during impaired auto-regulation) post-stroke hypertension remains unclear. Also the management of post-stroke hypertension is still a bone of contention. Some authors are convinced that prudent reduction of elevated blood pressure after stroke can actually improve cerebral vascular resistance (9). However, more recently the emphasis has changed (10).

Thus pre-stroke hypertension and immediate post-stroke hypertension seem to have differing aspects. This paper deals with neuro-epidemiologic investigation of these differing aspects.

MATERIAL AND METHODS

Neurologic epidemiology (descriptive, analytic and experimental) is the study of the distribution and determinants of neurologic diseases in human populations. Descriptive neuro-epidemiology characterizes observations of well-defined diseases in well-defined populations (11). The data of descriptive neuro-epidemiology frequently offer the possibility to formulate hypotheses concerning the etiology of disease. In a next step these hypotheses can be tested with a technique of analytic neuro-epidemiology (case-control study or prospective study).

The following data stem from the descriptive section of the first descriptive and analytic neuro-epidemiological study of stroke in The Netherlands. Descriptive data of the first two years of the study and analytic data of the case-control study are published (12-15). The study ws conducted in Tilburg (152.063 inh.) from 1978 to 1981 (Tilburg Epidemiological Study of Stroke: TESS). The basic definitions, guide-lines and record forms for registration were modelled, for the most part, after those of the WHO Multicenter Stroke Register (16). In order to be entered into the register, the patient had to fulfil the clinical criteria of stroke, officially reside in the city of Tilburg, and have the attack during the study period. All stroke patients were registered. Case-finding involved all existing sources (12).

During the study period 737 strokes occurred (400 in females, 337 in males). Eighty-five percent of the patients were admitted to the one existing neurological department in the city. Eighty percent of these patients were admitted on the first day after stroke. In the acute phase of the illness all stroke cases were examined by a physician and eighty-seven percent were examined by a neurologist. The average annual incidence of stroke per 100.000 population was 162. Cerebral infarction comprised the vast majority of strokes (eighty-two percent), intracranial hemorrhage contributed an additional fifteen percent, and three percent of strokes were not specified as to type.

To define hypertension we applied the WHO criteria (fifth Korotkoff phase) for use in epidemiological studies: normal range equal to or below 140/90, hypertensive range 160/95 and above, borderline or intermittent 140-159/90-94 mm Hg (17). Pre-stroke hypertension data stem from the patients' histories and the questionnaires filled out by family physicians and nursing home physicians who, before the start of the study, had been trained for collection of data. Post-stroke hypertension data stem from the clinical records (readings measured within 24 hours after admission and then daily). Statistical hypothesis testing for 2x2 tables employed both the Mantel-Haenszel tet (X_{mh}) and Fisher's "Exact Test" for comparison of two rates. For m x n contingency tables (with sufficient cells bearing expected frequencies of 5 or more), the chi-square (X^2) test was applied. Test results yielding 2 tail p-values of .05 less were considered statistically significant (13).

RESULTS

Pre-stroke hypertension

Fifty-nine percent of patients had a history of pre-stroke hypertension. Age was related to such occurrence (X^2 = 9.37, d.f.= 2, p = 0.01): in the age-group under 65 years 49 %, between 65 and 74 years 64 %, over 75 years 59 %. Females (67 %) suffered significantly more from

pre-stroke hypertension than males (48 %), even after adjusting for age
(Xmh = 5.00, p $<$ 0.0001). Especially in the age-group over 75 years the
difference was striking: 69 % among females and 42 % among males (table 1).
Hardly any difference in percentages of pre-stroke hypertension was noted
among the major types of stroke: in thrombo-embolic infarction 59 % and in
intracerebral hemorrhage 57 %. Only in the age-group below 65 years the
percentage of pre-stroke hypertension was higher among hemorrhage cases (56
%) than among infarction cases (48 %).

Post-stroke hypertension

Shortly after their stroke, 93 % of assessed cases were measured
hypertensive with systolic readings over 140 mm Hg and/or diastolic
readings over 90 mm Hg. Only 7 % had a normal post-stroke blood pressure.
The blood pressures of those over 65 years of age were significantly
higher, on the average, than of those younger (x^2 = 16.85, d.f.= 3,
p $<$ 0.001). Females had a higher proportion of definite hypertension, in
particular severe hypertension, than did males, regardless of age-adjusting
(XmH=3.06, p=0.002) (table 2). Post-stroke systolic blood pressures
demonstrated a similar proportion of post-stroke hypertension as that for
the joint measures of systolic and diastolic readings. Post-stroke
hypertension was found less frequently when viewing diastolic pressures
than when considering the combined systolic and diastolic ones (exact
p $<$ 0.0001). Furthermore, when only systolic blood pressures were compared,
females showed higher determinations than males did (x^2 = 14.30, d.f.= 3,
p=0.003). The overall distributions of combined systolic and diastolic
blood pressure did not differ significantly per stroke type. No association
was observed between stroke type and level of systolic pressure. However,
regardless of age, relatively more frequent severe hypertensive diastolic
readings were found in cases of hemorrhage than in cases of infarction (Xmh
= 2.79, p=0.01).

DISCUSSION

In the total group of strokes in this study pre-stroke hypertension
was noted in 59 % and post-stroke hypertension in 93 % of patients. In the
Comprehensive Stroke Center Data Base in North Carolina, Oregon and Monroe
County (New York), Toole and co-workers found the same 59 % of pre-stroke

Table 1   <u>PRE-STROKE HYPERTENSION AMONG STROKE CASES BY AGE AND SEX</u>

SEX/AGE	N	Arterial hypertension N	Arterial hypertension %	Treated immediately prior to stroke N	Treated immediately prior to stroke %	hypertensives under treatment %
Males						
< 65 yrs	101	44	44	38	35	80
65 - 74	123	72	59	60	49	83
75 +	113	47	42	33	29	70
Total	337	163	48	128	38	79
Females						
< 65 yrs	68	39	57	36	53	92
65 - 75	127	88	69	78	61	89
75 +	205	142	69	119	58	84
Total	400	269	67	233	58	87
Males & females						
< 65 yrs	169	83	49	71	42	86
65 - 74	250	160	64	138	55	86
75 +	318	189	59	152	48	80
Total	737	432	59	361	49	84

Table 2   <u>SYSTOLIC/DIASTOLIC BLOOD PRESSURE (mm Hg)</u>
<u>COMBINATION AFTER STROKE ONSET (ACUTE PHASE)</u>

	Males 313 N	Males 313 %	Males 313 %**	Females 363 N	Females 363 %	Females 363 %**	Total* 676 N	Total* 676 %
Normal (<140 and/or< 90)	29	9	9	17	5	5	46	7
Borderline hypertension (140-159 and/or 90-94)	66	21	22	55	15	15	121	18
Moderate hypertension (160-199 and/or 95-114)	131	42	44	151	42	41	282	42
Severe hypertension (200+ and/or 115+)	87	28	27	140	39	38	227	34

  * 61 cases were not assessed
** Age-adjusted to age distribution of total cases

	AGE GROUP <65 yrs 158 N	<65 yrs 158 %	65-74 yrs 235 N	65-74 yrs 235 %	75+ yrs 283 N	75+ yrs 283 %
Normal (<140 and/or <90)	19	12	14	6	13	5
Borderline hypertension (140-159 and/or 90-94)	38	24	31	13	52	18
Moderate hypertension (160-199 and/or 95-114)	52	33	102	43	128	45
Severe Hypertension (200+ and/or 115+)	49	31	88	37	90	32

hypertension in 4.125 stroke cases (18). Wallace and Levy found in 334 consecutive admissions for acute stroke elevation of blood pressure in 84 % on the day of admission. The blood pressure decreased spontaneously in the days following the acute event without specific antihypertensive therapy (19). The same gradual decrease was observed among 100 stroke patients by de Faire et al. (20) and by Melamed and co-workers (21). Melamed suggests that transitory increases of blood pressure after stroke may be due to general stress or to dysruption of central nervous system regulatory mechanisms. Feibel and co-workers noted a marked elevation of catecholamine excretion in 7 cases of acute post-stroke hypertension, all refractory to control by several antihypertensive drugs. In their opinion the intense sympathetic nervous system discharge resulting in acute refractory hypertension may be due to injury of the diencephalon or brainstem (or both) or to diffuse brain dysfunction from increased intracranial pressure or from intracranial blood (22).

In conclusion, neuro-epidemiology data indicate that pre-stroke hypertension and post-stroke hypertension differ quantitatively. There is also some suggestive evidence that their patho-physiology is different. According to the phraseology of Hughlings Jackson one may define pre-stroke hypertension as a negative condition and post-stroke hypertension as a positive condition. Pre-stroke hypertension, as a consequence and a source of atherosclerosis, is the most powerful risk factor for stroke and needs treatment without restrictions. Post-stroke hypertension, as a phenomenon of adaptation, protects the endangered cerebral blood flow during the totally pressure-dependent phase of disturbed cerebral auto-regulation.

ACKNOWLEDGEMENT

The Tilburg Epidemiological Study of Stroke was supported by grants from the Netherlands Heart Foundation and from the Netherlands Fund for Preventive Medicine.

REFERENCES

1. Wepfer, J.J. Oberservationes anatomicae ex cadaveribus eorum, quos sustulit apoplexia, cum exercitatione de eius loco affecto. Schaffhusii (J.C. Suteri), 1658.

2. McHenry L.C., Garrisons's history of neurology , Springfield (Charles C. Thomas) p. 81, 1969.

3. Kannel, W.B., Wolf, P.A., Verter, J., McNamara P.M. JAMA 214: 301-310, 1970.

4. Veterans administration cooperative study group: effects of treatment on morbidity in hypertension. I JAMA 202: 1028-1034, 1967.

5. Veterans administration cooperative study group: effects of treatment on morbidity in hypertension. II JAMA 213: 1143-1152, 1970.

6. Hypertension detection and follow-up program cooperative group. JAMA 242: 2562-2577, 1979.

7. The Oslo study, treatment of mild hypertension: a five year controlled drug trial. Am. J. Med. 69: 725-732, 1980.

8. The management committee. The Australian therapeutic trial in mild hypertension. The Lancet 1: 1261-1267, 1980.

9. Meyer, J.S., Sawada, T., Kitamura, A. et al. Neurology 18: 772-781, 1968.

10. Wise, R.J.S., Bernadi, S., Frackowiak, R.S.J., Legg, N.J., Jones, T., Brain 106: 197-222, 1983.

11. Schoenberg, B.S Neuroepidemiology 1: 1-16, 1982.

12. Herman, B., Schulte, B.P.M., Luijk, J.H. van, Leyten, A.C.M., Frenken, C.W.G.M. Stroke 11: 162-165, 1980.

13. Herman, B., Leyten, A.C.M., Luijk, J.H. van, Frenken, C.W.G.M., Op de Coul, A.A.W., Schulte, B.P.M. Stroke 13: 629-634, 1982.

14. Herman, B., Leyten A.C.M., Luijk, J.J. van, Frenken, C.W.G.M., Op de Coul, A.A.W., Schulte, B.P.M. Stroke 13: 334-339, 1982.

15. Herman, B., Schmitz, P.I.M., Leyten, A.C.M., Luijk, J.H. van, Frenken, C.W.G.M., Op de Coul, A.A.W., Schulte, B.P.M. Am. J. Epidemiol. 118: 514-525, 1983.

16. Hatano, S. Control of stroke in the community-methodological considerations and protocol of WHO stroke register (WHO document no. CVD/S/73.6 Rev. 1). Geneva, 1973.

17. Swales, J.D. Clinical hypertension, London (Chapman and Hall) p. 89, 1979.

18. Toole, J.F., Feibel, J., Yatsu, F.M., Becker, C., McLeroy, K.R., Howard, G., Coull, B., Walker, M.D. In: Abstracts 2nd International Symposium on Brain-Heart Relationship, Jerusalem, 1983.

19. Wallace, J.D., Levy, L.L. JAMA 246: 2177-2180, 1981.

20. Faire, U. de, Britton, M., Helmers, C., Wester, P.O.   Acta med. scand. __88__: 621-627, 1978.
21. Melamed, E., Cooper, G., Saltstein, E., Globus, M.   In: Abstracts 2nd International Symposium on Brain-Heart Relationship, Jerusalem, 1983.
22. Feibel, J.H., Baldwin, C.A., Joynt, R.J.   Ann. Neurol. __9__: 340-343, 1981.

11

THE RELATIONSHIP BETWEEN PLASMA RENIN ACTIVITY AND THE DEGREE OF IMPAIRED
CONSCIOUSNESS IN SPONTANEOUS SUBARACHNOID HEMORRHAGE

HAMANN, G., STOBER, T., BIRO, G., ANSTÄTT, T., SCHIMRIGK, K.

Department of Neurology, University of the Saarland, 6650 Homburg-Saar, FRG

ABSTRACT

The relationship between plasma renin activity and clinical
neurological deficits, especially signs of midbrain syndrom, was examined
in ten patients with spontaneous subarachnoid hemorrhage. The patients'
daily neurological examination was undertaken with particular attention
being paid to alterations in the degree of impaired consciousness, graded
according to the Glasgow Coma Scale. Plasma renin activity was determined
using an Angiotensin-I-radioimmunoassay on days 1-5;7;9;11;15;21 after the
bleeding episode. Six of seven patients who suffered from a deterioration
in the degree of consciousness, displayed a marked increase in plasma renin
activity. A negative correlation between plasma renin activity and the
values of the Glasgow Coma Scale resulted. The increase in plasma renin
activity could be caused by an increase in sympathetic activity as a result
of lesions in the midbrain or hypothalamus, resulting from elevated
intracranial pressure, intracerebral bleeding or vasospastic infarctions.

INTRODUCTION

The prognostic significance of plasma renin activity in subarachnoid
hemorrhage has already been postulated in 1980 (1). The aim of this
prospective clinical study was thus to determine the relationship between
the time course of plasma renin activity and neurological deficits after
spontaneous subarachnoid hemorrhage.

PATIENTS AND METHODS

The mean age of the ten patients we examined was 51.9 $\pm$ 10.7 years.
Nine of these patients suffered from hypertension, four of whom were never

treated. Four patients had a history of cardiac insufficiency, one of them had a myocardial infarction, and another one a coronary heart disease. All patients were examined serially. Specific attention was payed to signs of midbrain syndrom such as respiratory abnormalities, pupil motor changes, disturbances of eye movements and skeletal muscle motor function (2). The degree of impaired consciousness was assessed using the criteria of the Glasgow Coma Scale (3).

Subarachnoid hemorrhage was verified by means of cranial computed tomography and/or lumbar puncture. Panangiography with oblique projections was made of all patients to localize the aneurysm. Blood pressure and pulse rate were measured at least every two hours. An elevated pressure of more than 160 mmHg systolic and 95 mmHg diastolic was treated with clonidin and dihydralazin. The plasma renin activity was measured on days 1-5;7;9;11;15;21 after the subarachnoid hemorrhage using an Angiotensin-I-radioimmunoassay (4). The inter-assay error was 9.3 % and the intra-assay error 5.9 %. Disturbing influences were markedly reduced by drawing blood at the same time of day, through rapid transport in an icebath and immediate processing of the samples.

RESULTS

Seven patients showed signs of midbrain syndrom; of these six displayed a distinct concomitant elevation in plasma renin activity. The seventh patient suddenly developed a midbrain syndrom after a rebleeding episode, but without a subsequent increase in plasma renin activity. Furthermore, there was a significant negative correlation between changes of the Glasgow Coma Scale and plasma renin activity. An impairment of consciousness was accompanied by an increase in plasma renin activity, whilst an improvement in the degree of consciousness was associated with a decrease in the plasma renin level. There was no relationship between plasma renin activity and focal neurological findings. The expected correlation (5) between plasma renin activity and blood pressure was not present, whereby the use of antihypertensive drugs must be regarded as an influencing factor. The mean value of the plasma renin activity of the patient group was significantly higher (4.5 $\pm$ 4.1 ng/ml/h) than that of a control group (2-3 ng/ml/h) (Table 1). Those patients with cardiac insufficiency, and in particular the two patients with coronary heart disease or myocardial infarction had considerably higher values. The plasma renin values were always elevated well above the previous level in those five patients who died.

Table 1   TIME COURSE OF PLASMA RENIN ACTIVITY
            AND GLASGOW COMA SCALE VALUES OF THE 10 PATIENTS

patient No.	day after SAH	PRA ng/ml/h	GCS	patient No.	day after SAH	PRA ng/ml/h	GCS
1	1	0,03	14	6	1	0,36	14
	2	6,83	14		2	5,25	13
	3	3,05	5		3	1,62	14
	4	2,16	5				
	5	1,30	5	7	1	0,15	14
	7	1,73	6		2	0,14	13
	10	0,66	6		3	0,03	12
	11	0,40	6		4	0,30	12
	14	8,63	5		5	0,03	12
					7	0,98	0
2	0	0,98	13				
	1	2,56	6	8	1	2,60	14
	2	4,17	6		3	7,74	14
	3	2,85	6,5		6	6,43	13
	4	1,83	7		8	4,51	9
	6	4,85	5		9	2,55	9
	8	2,30	10		11	5,14	14
	10	6,91	3		15	0,09	14
3	2	0,73	13	9	1	0,85	5
	3	1,17	13		2	2,75	5
	4	0,51	12		3	1,02	5
	6	0,24	14		4	3,24	5
	8	0,21	14		5	5,64	3
	11	2,16	8		8	12,80	3
	12	2,73	14		9	2,39	5
	16	0,66	14		11	3,59	5
	22	2,55	14		13	7,17	9
					15	10,76	11
4	2	5,86	6		18	10,41	11
	3	8,33	3		22	8,88	11
5	3	1,76	14	10	3	1,26	5
	5	9,78	3		4	1,02	4
					6	2,39	3
					7	9,39	3
					8	11,44	3
					10	19,29	9
					12	10,07	11
					14	23,39	10
					22	-	0

SAH: subarachnoid hemorrhage
PRA: plasma renin activity
GCS: Glasgow Coma Scale

CASUISTRY

PATIENT 1

This 48 year old male patient was admitted one day after a subarachnoid hemorrhage; he was well orientated and conscious, with slight meningism and anisocoria. An angiography revealed an aneurysm of the anterior communicans arteria. He showed an increase in plasma renin activity and an accompanying deterioration on the Glasgow Coma Scale on the second day. After a temporary recovery, a renewed deterioration on the sixth day resulted in signs of midbrain syndrom accompanied by a subsequent rise in plasma renin activity. The patient died in a comatose state on the 17th day from pulmonary embolism and central cardiac insufficiency displaying a pronounced elevation in plasma renin activity shortly before his death (Fig. 1).

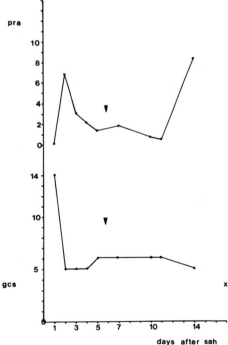

Fig. 1. Time course of plasma renin activity and Glasgow Coma Scale values in patient 1.

pra = plasma renin activity   gcs = Glasgow Coma Scale

▼ = midbrain syndrome   ✗ = death

PATIENT 2

This 62 year old female patient was hospitalized with an extreme headache, nausea and vomiting. Her blood pressure was 240/130 mmHg. Slight meningism and a slightly somnolent consciousness were prevalent. An angiography revealed an aneurysm in the anterior communicans arteria as cause of the subarachnoid hemorhage, which was particularly pronounced in the basal cysterns. The patient showed signs of midbrain syndrome on the third day after the subarachnoid hemorrhage with a significant increase in plasma renin activity. She died from cardiac insufficiency on the 12th day (Fig. 2).

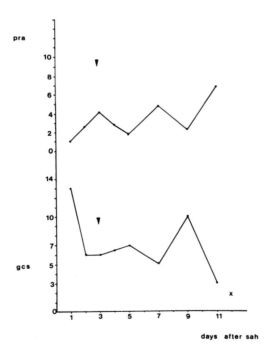

Fig. 2. Time course of plasma renin activity and Glasgow Coma Scale values
        in patient 2.
        pra = plasma renin activity   gcs = Glasgow Coma Scale
        ▼ = midbrain syndrome   ✕ = death

PATIENT 7

This 46 year old female patient was hospitalized in a comatose state with a
blood pressure of 220/120 mmHg. Excluding meningism, no focal neurological
deficit was prevalent. An aneurysm of the anterior communicans arteria was
demonstrated angiographically. On the third day after the subarachnoid
hemorrhage signs of midbrain syndrom developed, accompanied by a strong
elevation in plasma renin activity, which decreased as the clinical
situation improved subsequent to the ninth day (Fig. 3).

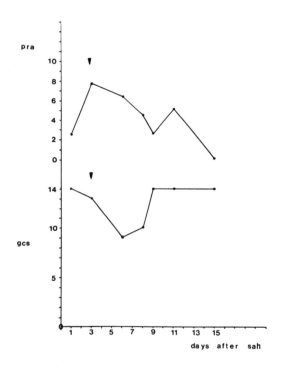

Fig. 3. Time course of plasma renin activity and Glasgow Coma Scale values
in patient 7.
pra = plasma renin activity   gcs = Glasgow Coma Scale
▼ = midbrain syndrome

PATIENT 9

This 56 year old male patient was soporous and right hemiparetic upcn
admission on the second day after a subarachnoid hemorrhage, caused by an
aneurysm of the anterior communicans arteria. After a short improvement he
developed signs of midbrain syndrom on the sixth day accompanied by an
elevated level of plasma renin activity. The pronounced rise in plasma
renin activity on the 12th day was concomitant with the development of a
non-resorptive hydrocephalus (Fig. 4).

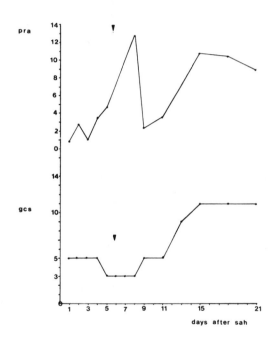

Fig. 4. Time course of plasma renin activity and Glasgow Coma Scale values
in patient 9.
pra = plasma renin activty   gcs = Glasgow Coma Scale
    = midbrain syndrome

DISCUSSION

The course of plasma renin activity subsequent to subarachnoid hemorrhage confirms the postulated prognostic significance of this parameter. Furthermore, it also allows pathogenic conclusions. The displayed relationship between elevated plasma renin activity and the development of midbrain syndrom, as well as the correlation between the degree of impaired consciousness and the level of plasma renin activity allow the assumption of damaged autonomic cardiovascular regulatory centers (6).

Midbrain syndrom (2) is characterized by impaired consciousness, but also substantially by disturbances in eye movement, skeletal muscle motor function and respiratory abnormalities, and can be caused by subarachnoid hemorrhage as a result of an elevation of intracranial pressure with ensuing dislocation of the brainstem, intracerebral bleeding or vasospastic infarction in this area. Besides midbrain structures, hypothalamic areas are also regularly affected, as the blood flow to the hypothalamus is supplied by small arterial branches, which originate from the large arteries of the Willis circle, which represent the most common localisation of berry aneurysms.

Our clinical results allow no further differentiation between these damage mechanisms, even not with regard to the participation of the central renin angiotensin system (7). Further pathological anatomical investigations are required to determine the pathogenic mechanisms and the extent to which the central renin angiotensin system is affected by these secondary lesions. As other signs of an elevation in sympathetic activity, such as an increase in the catecholamines had been observed in our patients (8), a sympathetically modulated rise in the plasma renin activity can be assumed (9,10).

For clinical purposes a prognosis can be just as accurately estimated from clinical parameters and/or from blood distribution and quantity on a cranial computed tomogram (11) as from the determination of plasma renin activty. As the latter is costly and time consumptive, it is unlikely to be of great clinical value.

REFERENCES

1. Neil-Dwyer, G., Walter, P., Shaw, H.J.H., Doshi, R., Hodge, M. Neurosurgery 7: 578-582, 1980.
2. McNealy, D.E., Plum, F. Archs Neurol. Vol 7: 26-47, 1962.
3. Teasdale, G., Jennet, B. Lancet I: 81-83, 1974.
4. Goodfriend, T.L In: Radioimmunoassay of angiotensins and renin activity in pepetide hormones (Ed. S.A. Berson) Yalow, New York, Elsevier, pp. 1158-1168.
5. Brunner, H.R., Laragh, J., Baer, L., Newton, M.A., Goodwin, F.T. Krakoff, L.R., Bard, R.H., Bühler, F.R. New. Engl. J. Med. 286: 441-449, 1972.
6. Doshi, R., Neil-Dwyer, G. J. Neurol. Neurosurg. Psychiat. 40: 821-826, 1977.
7. Ganten, D., Minnich, J.L., Hayduk, P.G.K., Brecht, H.M., Barbeau, A., Genest, R.B.J. Science 173: 64-65, 1971.
8. unpublished data
9. Neil-Dwyer, G., Cruickshank, J.M. J. Neurol. Sci. 23: 463-471, 1974.
10. Neil-Dwyer, G., Cruickshank, J.M., Stott, A., Brice, J. J. Neurol. Sci. 22: 375-382, 1974.
11. Turnbull, I.W. Br. J. Radiol. 53: 416-420, 1980.

This work is part of the dissertation of G. Hamann.

12

CONTINUOUS ELECTROCARDIOGRAPHIC MONITORING IN SPONTANEOUS SUBARACHNOID HEMORRHAGE

DI PASQUALE, G.(1), PINELLI, G.(1), ANDREOLI, A.(2), MANINI, G.L.(1), LA VECCHIA, F.(1), LUSA, A.M.(1), TENCATI, R.(1), TOGNETTI, F.(2)

(1) Service of Cardiology and (2) Department of Neurosurgery, Bellaria Hospital, Bologna, Italy

ABSTRACT

It is well known that intracranial hemorrhage, especially subarachnoid hemorrhage (SAH) may secondarily cause cardiac arrhythmias. In order to assess the incidence and the severity of arrhythmias, 85 consecutive patients aged 16-70 years (mean age 52 years) with SAH secondary to ruptured aneurysm were investigated with 24-hour Holter recordings.
Bradyarrhythmias and tachyarrhythmias were found in 76 patients (89%): premature ventricular beats in 43, ventricular tachycardia in 4, premature supraventricular beats in 25, paroxysmal supraventricular tachycardia and atrial fibrillation in 9, sinoatrial blocks and arrests in 25, atrioventricular dissociation in 3 and idioventricular rhythm in 2. Moreover, in 6 patients ST segment changes were found suggestive of transitory acute myocardial ischemia.
The presence and the severity of arrhythmias were correlated with the time elapsed from the episode of bleeding, with the QT interval duration.
Our results indicate a high incidence of arrhythmias in SAH, sometimes serious mainly in early stage. Continuous electrocardiographic monitoring is therefore extremely useful and provides data for therapeutic consideration.

INTRODUCTION

The association of cerebrovascular accidents, especially intracranial hemorrhage, with electrocardiographic ST-T abnormalities has been well known since three decades (1-4). Conversely, the association of acute intracranial events, particularly subarachnoid hemorrhage (SAH), with cardiac arrhythmias has been recognized only in recent years and few studies of electrocardiographic monitoring are available (5-10).

In order to assess the incidence and severity of cardiac arrhythmias in SAH, we carried out a prospective study with 24-hour Holter recording.

MATERIAL AND METHODS

We studied 85 consecutive patients (43 males and 42 females) aged 16-70 years (mean age 52 years) admitted to the Neurosurgery Department with SAH secondary to ruptured intracranial aneurysm.

Thirtyseven of the 85 patients were admitted acutely, within 48 hours of the hemorrhage. According to the Hunt-Hess (11) grading 46 patients were grade 1-2 (awake, headache, neck stiffness), 23 grade 3 (drowsiness, confusion or mild focal deficits), 16 grade 4-5 (stupor to deep coma).

The diagnosis of SAH was made by the neurologist by means of physical examination, cerebral computed tomography and/or lumbar puncture. Cerebral angiography was performed in all patients except two, in order to visualize the origin of the hemorrhage (Table 1).

Table 1
ANGIOGRAPHIC FINDINGS IN 85 PATIENTS WITH SUBARACHNOID HEMORRHAGE

Angiographic Lesion	No. Cases
ICA aneurysm	16
MCA aneurysm	16
ACoA aneurysm	25
Posterior circulation aneurysm	3
No vascular malformation	17
Angiography not performed	2

ICA = internal carotid artery
MCA = middle cerebral artery
ACoA = anterior communicating artery

On admission each patient had a clinical cardiological evaluation, a standard electrocardiogram, chest X-ray and serum electrolytes determination. On the same day all patients underwent a 24-hour Holter recording. All Holter recordings were performed using Avionics 445 A two-channel recorders and the tape recordings were analyzed using an

Electrocardioscanner Avionics 660 B provided with arrhythmias computer. In every patient the QTc interval was measured, using the Bazet formula, on an average of ten complexes of a standard ECG performed on the same day as the Holter monitoring.

RESULTS

Cardiac arrhythmias were found in 76 patients (89%) and in 6 cases ST-segment changes were found suggestive of transitory acute myocardial ischemia (Table 2).

Premature ventricular beats (PVBs) were found in 43 patients and 21 of these had multiform and successive PVBs or R-on-T phenomenon (3rd-5th Lown classes). Four patients with frequent PVBs also had episodes of nonsustained ventricular tachycardia (Fig. 1).

Supraventricular arrhythmias were found in 34 cases: premature supraventricular beats in 25, paroxysmal supraventricular tachycardia in 7 and atrial fibrillation in 2.

Thirtysix patients had sinus bradycardia ($<$ 50 beats/min), 20 marked sinus tachycardia ($>$ 120/min), and 28 sinus arrhythmia. Hypokinetic arrhythmias were found in 30 patients: sinoatrial blocks in 21, sinus arrests 4" in 4, atrioventricular dissociation in 3 and idioventricular rhythm in 2 (Fig. 2). Three patients, grade 4-5 Hunt-Hess, died during Holter monitoring: the ECG showed asystole which followed periods of marked sinus tachycardia.

Finally, asymptomatic transitory ventricular repolarization abnormalities, suggestive of acute myocardial ischemia, were detected in 6 patients. They consisted of a ST-segment depression in lead CM5 of 10 to 30 minutes in 5 cases and of a ST-segment elevation of 20 minutes in 1 case. In one patient the episode of ST-segment depression occured during cerebral angiography and was associated with bigeminal PVBs.

The presence and severity of arrhythmias were correlated with the time elapsed from the episode of bleeding. Patients were divided into 2 groups: early (37 cases) and late stage (48 cases), whether the Holter monitoring was performed within or after 48 hours from the hemorrhage. The severity of arrhythmias was significantly higher in patients studied at an earlier stage of SAH.

The QTc interval was measured in 83 of the 85 patients, 2 cases with

Table 2

CARDIAC ARRHYTHMIAS AND TRANSITORY ST-T CHANGES DETECTED IN 85 PATIENTS WITH SUBARACHNOID HEMORRHACE.

TYPE OF ARRHYTHMIA	EARLY STAGE N = 37	LATE STAGE N = 48
PVB	34	9
Non sustained VT	4	-
SVPB	15	10
Paroxysmal AF	2	-
Paroxysmal SVT	4	3
Sinus bradycardia	20	16
Sinus tachycardia	18	2
Sinus arrhythmia	16	12
S-A blocks	14	7
S-A arrests $>$ 4"	3	1
A-V dissociation	3	-
Idioventricular rhythm	2	-
Asystole	3	-
ST-segment depression	4	1
ST-segment elevation	1	-

PVB = premature ventricular beat
VT = ventricular tachycardia
SVPB = supraventricular premature beat
AF = atrial fibrillation
SVT = supraventricular tachycardia
S-A = sinoatrial
A-V = atrio-ventricular

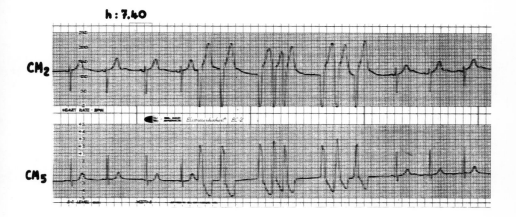

Fig. 1. Episode of nonsustained ventricular tachycardia.

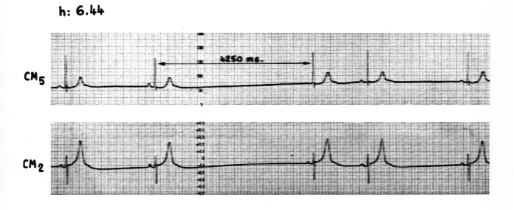

Fig. 2. Marked sinus bradycardia at heart rate of 30/min and sinus arrest
of 4.250 msec followed by escape atrial beat.

bundle branch block being excluded. A prolonged QT interval was present in 36 cases. No significant difference was found for the QT interval duration between the patients with and those without hyperkynetic ventricular arrhythmias, even if a prolonged QT interval was frequently observed in patients with serious ventricular arrhythmias.

DISCUSSION

The results of the present study indicate an extremely high incidence of arrhythmias in SAH, which may be severe particularly in the acute period. The majority of serious arrhythmias was in fact observed in patients studied within the first 48 hours from the hemorrhage. This is in agreement with scattered case reports published in the literature. Life-threatening arrhythmias, such as ventricular flutter or fibrillation and torsade de pointe have been in fact reported in patients admitted to the Hospital in the acute phase of SAH (12-14). It is therefore possible that serious ventricular arrhythmias are responsible for sudden death or for the initial loss of consciousness after SAH.

Several clinical series have verified the association of SAH with cardiac arrhythmias but only few prospective studies of electrocardiographic monitoring have been done (6,9,10). Estanol (6) found some kind of cardiac arrhythmias, including sinus bradycardia, in all the 15 patients with SAH monitored for at least 5 days in an intensive stroke care unit; 20% of patients had serious life-threatening arrhythmias. Mikolich (7) found complex PVBs in 15 of 30 patients studied by 24-hour Holter monitoring in the acute period following a thromboembolic or hemorrhagic cerebral accident.

Several mechanisms have been proposed to explain the origin of arrhythmias in SAH. Experimental studies demonstrated that sympathetic and vagal discharges were responsible for cardiac arrhythmias in acute intracranial diseases, particularly SAH (15). In the dog the introduction of blood into the subarachnoid space produces immediate arrhythmias closely correlated with the increase of intracranial pressure (7,16). On the other hand a chronic elevation of intracranial pressure usually induces only sinus bradycardia; no serious arrhythmias were in fact observed in our patients studied in a late stage. The sudden increase of intracranial pressure possibly triggers the sympathetic and vagal discharge because of a

169

compression of the brain stem or diencephalic structures (16,17). Moreover, many clinical and experimental data indicate an increase of circulating catecholamines in patients with SAH (17-19). The complex relationship between sympathetic and parasympathetic systems may determine the prevalence of tachy- or bradyarrhythmias or the bradycardia-tachycardia syndrome.

The correlation between QT interval prolongation and cardiac arrhythmias in SAH has not been clearly established. A prolonged QT interval is frequently observed in patients with SAH and serious ventricular arrhythmias associated with this electrocardiographic pattern are reported (6,13,14). The 42% of our patients had a prolonged QT interval, but this was not correlated with a higher incidence of ventricular arrhythmias.

In conclusion, our data demonstrate the importance of a continuous electrocardiographic monitoring in patients with SAH. A careful monitoring in the earlier stage after the bleeding allows the detection and treatment of serious cardiac arrhythmias.

REFERENCES

1. Abildskov, J.A., Millar, K., Burgess, M.J., Vincent, W. Prog. Cardiovasc. Dis. 13: 210-216, 1970.
2. Burch, G.E., Myers, R., Abildskov, J.A. Circulation 9: 719-723, 1954.
3. Byer, E., Ashmann, R., Toth, L.A. Am. Heart J. 31: 796-806, 1947.
4. Wong, T.C., Cooper, E.S. Am. J. Cardiol. 23: 473-477. 1969.
5. Britton, M., De Faire, V., Helmers, C., Miah, K., Ryding, C., Wester, P.O. Acta med. scand. 205: 425-428, 1979.
6. Estanol, B.V., Dergal, E.B., Cesarman, E., San Martin, O.M., Loyo, M.V., Vargas Lugo, B., Perez Ortega, R. Neurosurgery 5: 675-680, 1979.
7. Mikolich, J.R., Jacobs, W.C., Fletcher, G.F. JAMA 246: 1314-1317, 1981.
8. Srivastava, S.C., Robson, A.O. Lancet 2: 431-435, 1964.
9. Reinstein, L., Gracey, J.G., Kline, J.A., Van Buskirk, C., Arch. Phys Med Rehabil 53: 311-314, 1972.
10. Lavy, S., Yaar, J., Melamed, E., Stern, S. Stroke 5: 775-780, 1974.
11. Hunt, W.E., Hess, R.M. J. Neurosurg. 28: 14-20, 1968.

12. Estanol, B.V., Marin, O.S.M. Stroke 6: 382-386, 1975.
13. Carruth, J.E., Silverman, M.E. Chest 78: 886-888, 1980.
14. Hust, M.H., Nitsche, K., Hohnloser, S., Bohm, B., Just, H. Clin. Cardiol. 7: 44-48, 1984.
15. Hunt, D., McRae, C., Zapf, P. Am. Heart J. 77: 479-482, 1969.
16. Estanol, B.V., Loyo, M.V., Mateos, J.H., Foyo, E., Cornejo, A., Guevara, J. Stroke 8: 440-447, 1977.
17. Weidler, D.J. Stroke 5: 759-764, 1974.
18. Boddin M., Bogaert, A.V., Dierick, W. Cardiology 58: 229-237, 1973.
19. Feibel, J.H., Campbell, R.G., Joynt, R.J. Trans. Am. Neurol. Assoc. 101: 242-249, 1976.

13

PATHOLOGICAL-ANATOMICAL FINDINGS WITHIN THE AUTONOMIC REGULATORY CENTERS IN PATIENTS WITH NEUROGENIC CARDIOVASCULAR REGULATORY DYSFUNCTIONS

ANSTÄTT, T. (1),STOBER, T.(1), MERKEL, K.H.(2), SEN, S.(3), FREIER, G.(1), SCHIMRIGK, K.(1)

(1) Neurologische Klinik, (2) Pathologisches Institut, (3) III. Medizinische Klinik der Universitätskliniken 6650 Homburg/Saar, FRG

ABSTRACT

Autonomic regulatory disturbances due to cerebrovascular diseases were investigated in a prospective study of 147 patients with intracranial hemorrhage and cerebral infarction. These patients were neurologically examined on a daily basis along with 12 lead standard ECGs and long-term Holter ECGs. The brains of 14 patients who died during this study were examined for morphological alterations within the autonomic regulatory centers. Sections were made of the medulla oblongata at three different levels, through the pons, mesencephalon, and hypothalamus in three planes each, as well as from the temporal lobe within the nucleus amygdalae.

Pathological alterations of various degrees were found in the brain stem, diencephalon, and/or within the limbic system. These consisted of primary ischemic lesions and hemorrhages, as well as secondary hemorrhages in the brain stem following an increase of intracranial pressure, and secondary ischemic damage due to angiospasm after subarachnoid hemorrhage. Case studies of patients with cerebral infarction and subarachnoid bleeding are presented. it can be assumed that the histologically observed lesions are related to the clinical course and the occurance of ECG changes. Our findings demonstrate that damage to the autonomic regulatory centers can be considered the cause of cardiovascular dysfunction in cerebrovascular disease.

INTRODUCTION

It is known from numerous studies that ECG changes may occur after a variety of diseases of the central nervous system, especially subsequent to intracranial hemorrhage (1-9). This induced us to carry out a prospective long-term ECG study in patients with subarachnoid hemorrhage, intracerebral hemorrhage, and cerebral infarction. A total of 120 patients with intracranial hemorrhage and 30 patients with cerebral infarction have so far been investigated. Many of the patients showed ECG abnormalities, such as alterations of the P wave and of the S-T segment. Diverse patterns of cardiac arrhythmia were registrated in long-term ECG. We observed asystoles > 2s, supraventricular extrasystoles, supraventricular tachycardia, monotopic and polytopic ventricular extrasystoles, nonsustained ventricular tachycardia, and in two cases a polymorphic tachycardia type of torsade de pointes.

On the basis of animal experiments and neuroanatomical studies (10-12) which in the past increased our knowledge of autonomic regulatory centers in the brain stem, diencephalon and structures of the limbic system, especially the amygdaloid nucleus, the hypothesis could be made that lesions of these autonomic regulatory centers associated with cerebrovascular disease are responsible for the observed ECG changes. In order to verify this hypothesis, we started a series of pathological-anatomical examinations of patients from our prospective ECG-study. The purpose was to morphologically identify lesions in the brain stem, diencephalon or in structures of the limbic system.

PATIENTS AND METHODS

4 patients with subarachnoid hemorrhage, 3 patients with intracerebral hemorrhage and 7 patients with cerebral infarction have so far been examined by clinical, as well as pathological-anatomical methods (Table 1).

These patients were neurologically examined on a daily basis along with daily regular standard ECG (12 leads). Long-term Holter ECGs were carried out over an average of 5 days. ECG changes were considered neurogenic if an unambiguous correlation between the alterations in the ECG pattern and a change in the clinical neurological findings existed. The brains were examined for morphological alterations within the autonomic regulatory centers. Sections were made of the medulla oblongata at three different levels, through the pons, mesencephalon, and hypothalamus in three planes each, as well as from the temporal lobe within the nucleus amygdalae.

Table 1   PATHOLOGICAL-ANATOMICAL FINDINGS

No.	Diagnosis	Age/Sex	Cause of Death	Survival Time (Days)	Neurogenic ECG Alterations	N. Amygdalae (Limbic System)	Hypothalamus	Mesencephalon	Pons	Medulla oblongata
1	SAH	47/m	RB (?), TTH	34	+	I	I		I	
2	SAH	44/f	RB, TTH	13	+		I,H			
3	SAH	48/m	RB, TTH	17	+	I (Fornix)	I			I
4	SAH	42/f	TTH	3		I (Fornix)			H	H
5	ICH	85/m	pulm. embolism	22	+			H	H	H
6	ICH	55/m	TTH	1						
7	ICH	63/f	cerebral hem.	1	+				H,I	H,I
8	Inf	78/m	?	6		I	I			
9	Inf	78/m	brain stem inf	6	+		I	I	I	
10	Inf	42/m	TTH	4	+			H	H	
11	Inf	61/m	brain stem inf	1		I		I,H	I	
12	Inf	24/m	TTH	1						
13	Inf	72/m	TTH	2		I	I	H	H	
14	Inf	35/m	TTH	4		I (Fornix)	I			

Abbreviations:

SAH = subarachnoid hemorrhage; ICH = intracerebral hemorrhage; Inf = cerebral infarction

TTH = transtentorial herniation; RB = rebleeding; I = infarction; H = hemorrhage

RESULTS

Neurogenic ECG-alterations were observed in 3 of 4 patients with subarachnoid hemorrhage, in 2 of 3 patients with intracerebral hemorrhage and in 2 of 7 patients with cerebral infarction (Table 1).

Pathological alterations of various degrees were found in the brain stem, diencephalon, and/or within the limbic system in all but two patients. These consisted of ischemic lesions, as well as hemorrhages.

We shall compare the clinical course, the ECG findings, and the pathological-anatomical findings of 1 patient with a cerebral infarction and a second patient with a subarachnoid hemorrhage.

1. A 42-year old male patient (No. 10) suffered from a cerebral infarction subsequent to an occlusion of the left internal carotid artery. This was followed two days later by an increase of the intracranial pressure, which lead to respiratory insufficiency requiring intubation and artificial respiration. Subsequent to this, both pupils were fixed and dilated. At this time, computerized tomography revealed a very extensive midline shift to the right. Finally, a decrease in blood pressure developed which could not be sufficiently controlled by medication. The patient died 4 days after the cerebral infarction.

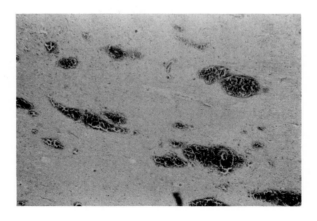

Figure 1. Hemorrhages without glial reaction in the mesencephalon of patient No. 10 with a cerebral infarction.

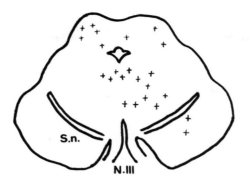

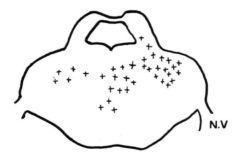

+ = hemorrhage

Figure 2. Localisation and extension of hemorrhages in the brainstem of
patient No. 10 with a cerebral infarction. Sections in the level
of the mesencephalon and of the pons.

During the increase of intracranial pressure, long-term ECG revealed a
number of polytopic ventricular extrasystoles (up to 20/min) as well as
nonsustained ventricular tachycardia. At that time, the QTC interval was
prolonged; before and after this phase it was within the normal range.
Autopsy revealed a recent cerebral infarction in the area supplied by the
left medial and anterior cerebral arteries as well as generalized cerebral

edema with uncal herniaton and cerebellar pressure cone. Histological examination confirmed the recent cerebral infarction in the above described area. The hypothalamus and the amygdaloid nucleus were not affected. There were, however, hemorrhages into the mesencephalon as well as into the pons. The localisation of these hemorrhages in the upper brainstem - verified by histological examination (Figure 1) - is shown in Figure 2. There were no lesions in the medulla oblogata.

2. A 47-year old male patient (No. 1) was hospitalized subsequent to a subarachnoid hemorrhage. There was a history of hypertension, but none of cardiac disease. From the beginning of the hospitalization, the patient varied from soporous to comatose. He always demonstrated disturbances of the ocular and pupillary motor responses indicating a brain stem lesion. These disturbances were most pronounced 10 days after hospitalization and were then associated with nonsustained ventricular tachycardia as seen in the long-term ECG. After a temporary improvement, loss of the pupillary and corneal reflexes - as an indication of secondary brainstem lesion - reoccurred 5 weeks after the hemorrhage. The patient finally died of central cardiovascular failure subsequent to this complication.

Autopsy revealed an aneurysm of the anterior communicating artery as the source of hemorrhage. The subarachnoid hemorrhage was most pronounced in the basal cisterns. There was a massive cerebral edema. Signs of tentorial herniation as well as herniation into the foramen magnum were evident. A recent ischemic necrosis could be seen macroscopically in the left frontral lobe.

A section through the hypothalamus at the level of the optic chiasm showed circumscribed infarctions with a glial reaction in the areas of the corpus callosum, the fornix and the hypothalamus (Figure 3). The preoptic nucleus and the supraoptic nucleus were also afflicted. The next section at the level of the infundibulum also revealed an infarction of the hypothalamus affecting the paraventricular nucleus. In addition, an extensive ischemic necrosis could be identified in the amygdaloid nucleus. At the level of the corpora mamillaria, there were multiple infarctions of similar histological appearance in the caudate, thalamus and hypothalamus as well as in the hippocampal cortex. Some of these infarctions did not however show a glial reaction. They probably developed after a rebleeding a short time before death. Histological analysis of the brain stem revealed an ischemic necrosis with glial reaction in the central pons (Figure 4).

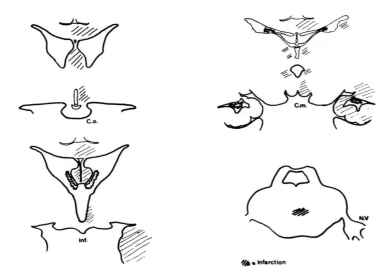

Figure 3. Localisation and extension of infarctions in patient No. 1 with
a subarachnoid hemorrhage. Sections in the level of the optic
chiasm (C.o.), Infundibulum (Inf.), mamillary bodies (C.M.), and
of the pons.

♥ : infarction
borderline

Figure 4. Histological appearance of the infarction with glial reaction
in patient No. 1 with a subarachnoid hemorrhage.

DISCUSSION

Pathological-anatomical examination of the brainstem, hypothalamus, and of the limbic system of patients who have died of cerebrovascular disease has revealed hemorrhages and infarctions.

In general such lesions may be primary in origin, that is they may develop as part of the primary cerebrovascular event. It may however be assumed , that in most of our patients, the hemorrhages and infarctions were secondary in origin. Thus they may have been caused by one of two mechanisms:

1) By tentorial herniation and herniation into the foramen magnum caused by brain edema, which leads to compression of brain tissue and blood vessels, especially at the free tentorial edge and thus to infarctions and hemorrhages in basal parts of the temporal lobe and in the brain stem. This pathogenic mechanism may have occured in our first patient.

2) By angiospastic ischemia following subarachnoid hemorrhage as can be assumed in our second patient.

In the first patient with the cerebral infarction, the hemorrhages into the brainstem did not show a glial reaction. The symptomatology of the brainstem lesion which developed one day prior to death, and the concomitant neurogenic ECG changes may thus well have originated from this secondary hemorrhage. Some of the infarctions in the second patient with a subarachnoid hemorrhage showed a pronounced glial reaction. Massive cardiac arrhythmia had been observed about three weeks prior to death. It is hence possible to claim a causal relationship between these infarctions and the occurrence of these ECG alterations. Those infarctions without glial reaction must have developed later. The electrocardiographic changes cannot be attributed to these alterations

Finally, it has to be considered whether or not the localisation of the observed lesions permits the conclusion that the ECG changes were caused by this tissue damage. It can be assumed that the secondary hemorrhages into the brainstem of our first patient resulted in a lesion of the following cerebral structures, which are involved in central autonomic regulation: in the level of the mesencephalon ascending and descending pathways between the hypothalamus and the nucleus amygdalae and autonomic regulatory centers in the lower brain stem may be damaged; in the pons there were additional lesions of the nucleus rhaphe magnus, the locus coeruleus and the Kölliker-Fuse nucleus (Figure 2). In our second patient the secondary angiospastic infarctions had damaged the columnae fornicis, the preoptic, supraoptic, paraventricular and posterior nucleus of the hypothalamus, one mamillary body, the hippocampus, and the nucleus rhaphe magnus (Figure 3).

It can be concluded that in patients with cerebrovascular disease primary and secondary infarctions and hemorrhages into autonomic regulatory centers may be responsible for the occurrence of cardiac dysfunction. This assumption is substantiated both by the histological age and by the localisation of the observed tissue destructions in our patients.

REFERENCES

1. Andersson, G.J., Woodburn, R., Fish, C.   Am. Heart J. 86: 395-398, 1973.
2. Aschenbrenner, R.G., Bodechtel, G.   Klin Wschr 17: 298-302, 1983.
3. Anstätt, T., Stober, T., Sen, S., Burger, L., Rettig, G., Schimrigk, K. Intensivmed. 20: 90-94, 1983.
4. Ashby, D.W., Chadha, J.S.   Br. Heart J. 30: 732-734, 1968.
5. Estanol, B.V., Loyo, J.V., Mateos, J.H., Goyo, E., Cornejo, A., Guevard, J.   Stroke Vol. 8, No. 4: 440-447, 1977.
6. Hersch, C.   Br. Heart J. 83: 232-236, 1972.
7. Myers, M.G., Noris, J.W., Hachinski, V.C., Weingert, M.E., Sole, M.J., Stroke Vol. 13, No. 6: 838-842, 1982.
8. Smith, M., Ray, C.T.   Dis. Chest 61: 125-133, 1972.
9. Stober, T., Kunze, K.   J. Neurol. 227: 99-113, 1982.
10. Blessing, W.W., West, M.J., Chalmers, J.   Circulation Res. Vol. 49, No. 4: 949-958, 1981.
11. Loewy, A.D., McKellar, S.   Fed. Proc. 39: 2495-2503, 1981.
12. Stock, G., Schmelz, M., Knuepjer, M.M., Forssmann, W.G.   In: Current Topics in Neuroendocrinology. (Eds. D. Ganten, D. Pfaff), Springer, Berlin, Heidelberg, New York, Tokyo, 1983.

14

THE INSULA AND ALTERED SYMPATHETIC ACTIVITY AFTER EXPERIMENTAL STROKE

HACHINSKI, V.C., SMITH, K.E., CIRIELLO, J., GIBSON, C.J., SILVER, M.D.

Departments of Physiology, Clinical Neurological Sciences, and Pathology, Health Sciences Centre, University of Western Ontario, London, Ontario, Canada, N6A 5C1

ABSTRACT

Our previous studies in an intensive care stroke unit have demonstrated an increased incidence of cardiac arrhythmias, serum cardiac enzymes and plasma catecholamines in stroke patients compared to controls. In addition, most of the patients dying from stroke had myocardial lesions similar to those seen in patients with phaeochromocytoma and in animals after systemic infusion of catecholamines. Taken together, this evidence suggests altered sympathetic activity following human stroke. To investigate this problem, two series of experiments were done in an experimental stroke model in the cat: somatosympathetic reflexes and measurement of plasma catecholamines. The first series showed that the late component of the somatosympathetic reflex elicited by stimulation of the sciatic nerve and recorded from either the T2 or T3 white ramus, or the inferior cardiac nerve was significantly increased in amplitude after experimental stroke (Soc of Neuroscience abstract # 515 1984). The second series showed that plasma norepinepherine levels were significantly elevated in the experimental stroke animal as compared to the sham operated animal (Stroke 16: 136, 1985). In all experimental stroke animals, the extent of induced unilateral ischemia varied. However, it was observed that only the animals with involvement of the insular cortex showed alterations in sympathetic activity. These data suggest that the insula plays an important role in the changes in sympathetic activity seen after experimental stroke.

CLINICAL BACKGROUND

Patients recovering from stroke may perish unexpectedly from cardiac complications or sudden death (1). Systematic observations in an acute stroke unit established that cardiac arrhythmias occurred more commonly in stroke patients and that serum cardiac enzymes and plasma catecholamines were raised in these patients when compared to similarly managed controls, even when matched for age, sex and the presence of heart disease (2-5). Moreover, at autopsy many patients with stroke showed myocardial lesions (6) similar to those seen in animals infused with catecholamines (7,8). These data suggest that the brain lesion can lead to cardiac complications.

Although a relationship between intracranial lesions and cardiac abnormalities has long been documented (9,10) most of the work has dealt with raised intracranial pressure and intracranial hemorrhage (11-14) rather than with focal cerebral ischemia, the most common form of stroke. Consequently we undertook to study autonomic disturbances after experimental focal ischemia.

EXPERIMENTAL MODEL

We successfully adapted the well established stroke cat model of O'Brien and Waltz (1973) (15). Our working hypothesis was that altered sympathetic activity is responsible for most of the autonomic complications seen after focal cerebral ischemia. Three indices of sympathetic activity were measured:

1. Changes in somatosympathetic reflexes after stroke
2. Changes in myocardial tissue after acute cerebral infarction
3. Changes in plasma catecholamines

METHODS

Experiments were done in adult cats of either sex weighing 2.0 to 3.5 kg. In brief, the left middle cerebral artery (MCA) was isolated and an occluding device was placed around the MCA distal to the lentriculostriate arteries in animals under ketamine anesthesia. After a minimal period of 4-7 days, the animals were reanesthetized with alpha-chloralose and paralyzed and artifically ventilated. The femoral artery and vein were routinely cannulated for recording arterial pressure and administration of drugs, respectively. The sciatic (ScN) and superficial radial (SRN) nerves were isolated and placed on bipolar stimulating electrodes and the T2-T3 white rami were isolated and placed on recording electrodes. Stimulation and recording of electrical activity from the nerves was done using conventional electrophysiological methods. The MCA was permanently occluded by retracting the occluding device. In selected experiments for plasma catecholamine determinations after stroke, the animals underwent the same surgical procedures except for the isolation of the nerves.

RESULTS

Altered Somatosympathetic Reflexes

In 15 experimental and 9 control cats, the SRN and ScN were stimulated (0.5-1.5 mA, 2 ms, 1 Hz) evoking early and late responses in either the T2 or T3 white ramus with peak latency of 10-45 ms and 55-90 ms, respectively.

The stroke cats were separated into two groups based on the direction of change of the early and late components of the evoked responses to somatic nerve stimulation. As compared to initial values obtained from the same animals, occlusion of the MCA resulted in a significant decrease in the peak-to-peak amplitude of the early reflex responses and a significant increase in the peak-to-peak amplitude of the late reflex response during stimulation of the SRN and ScN. As well, the amplitude of the late component during SRN and ScN stimulation was significantly greater and the amplitude of the early component during SRN stimulation was significantly lower in stroke animals compared to that obtained from control animals. There was also a small group of animals in which somatosympathetic reflexes decreased in amplitude after stroke.

The early component of the somatosympathetic reflex follows a spinal pathway, segmentaly organized and restricted to near the level of the afferent input whereas the late component arises from a more general response giving through the medulla. It is likely that the early involvement of the response after stroke was inhibited locally, while the late components freed from cortical inhibition increased in amplitude reflecting enhanced sympathetic activity.

Acute Myocardial Damage

In 23 chloralosed, paralyzed and artifically ventilated cats we examined the effects on the myocardium of occluding (n=17) or sham-occluding (n=6) the left MCA.

Twelve to twenty-two hours after the MCA occlusion, the animals were perfused via the femoral artery with 1 liter of 0.9 % physiological saline followed by 1 liter of 10 % buffered formalin-carbon dye solution. The brain and heart of each animal was removed and placed in a 30 % sucrose-buffered formalin solution. The presence and extent of the cerebral infarction was determined by the extent of the unstained area.

Each heart was cut into three transverse sections, one near the apex, one in mid-position, and one near the base. The sections included the right and left ventricular walls and the septum. They were embedded in paraffin, sectioned at 50 um and stained with either hematoxylin and eosin or the Movat stain.

After occlusion of the MCA for 12-22 hr, 41 % (7/17) of the hearts had either acute myocardial necrosis (3/7), focal hemorrhage (3/7), or both (1/7) (Figure 1). None of the animals without raised catecholamines showed acute lesions.

The acute cardiac lesions can also be interpreted as reflecting increased sympathetic activity, as they resemble closely the myocardial changes described after infusion of catecholamines in experimental animals (7,8) and those induced by electrostimulation of the orbital frontal cortex (16) and the posterior hypothalamus (17) in the monkey.

Plasma norepinephrine (NE) and epinephrine (E) levels were significantly increased in animals with acute cardiac lesions over pre-stroke values and as compared to cats without acute myocardial changes and sham-operated cats. This suggests a direct role for raised catecholamine levels in myocardial damage.

Increased Plasma Catecholamine Levels and Insular Damage

Blood samples (3 ml) were taken from the cannulated femoral artery at timed intervals before and after the stroke or sham-stroke. The samples were centrifuged immediately, and the plasma collected and stored at -70° C until assayed for catecholamines. The concentrations of catecholamines (epinephrine, dopamine, and norepinephrine) were determined by high pressure liquid chromatography (HPLC) system. In animals with acute myocardial damage the levels of plasma NE and epinephrine (E) were significantly increased compared to pre-MCA occlusion values (+ 46 $\pm$ 18 %

and + 142 ± 45 %, respectively). As well, in cats with acute myocardial damage, changes from initial levels of plasma NE and E were significantly increased over those of experimental cats without acute myocardial damage. In animals which did not have acute myocardial damage (10/17) the circulating plasma levels of catecholamines were not significantly different from pre-MCA occlusion values. Similarly, sham-occlusion did not alter plasma catecholamine levels. Moreover, rises in plasma catecholamines only ocurred in animals with ischemic damage to the insula.

These data suggest an important role for the insula in mediating the cardiac complications of stroke and is in keeping with the known role and connections of the insula with centers subserving autonomic function (18-20).

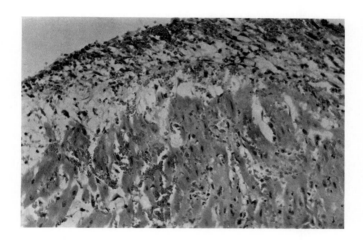

Figure 1: Brightfield photomicrograph of hematoxylin and eosin stained transverse section through the left ventricle showing acute myocardial damage in cats 12-22 hr after permanent occlusion of the MCA. The endothelium is uplifted and separated from the underlying tissues. Subendocardial hemorrhage as well as eosinophilic and vacuolated myocardial tissue are present. Calibration 100 um

REFERENCES

1. Silver, F.L., Norris, J.W., Lewis, A.J., Hachinski, V.C. Stroke 15: 492-496,1984.
2. Norris, J.W., Hachinski, V.C., Myers, M.G., Callow, J., Wong, T., Moore, R.W. Stroke 10:201-206,1979.
3. Myers, M.G., Norris, J.W., Hachinski, V.C., Sole, M.J. Stroke 12: 200-204,1981.
4. Myers, M.G., Norris, J.W., Hachinski, V.C., Weingert, M.E., Sole, M.J. Stroke 13:838-842,1982.
5. Hachinski, V.C., Norris, J.W. (1985) The acute stroke. Vol 27 Contemporary Neurology Series. F.A. Davis. Philadelphia pp.
6. Kolin, A., Norris, J.W. Stroke 15:990-993,1984.
7. Chappel, C.I., Rona, G., Balazs, T., Gaudry, R. Can. J. Biochem. Physiol. 37:35-42,1959.
8. Csapo, Z., Dusek, J., Rona, G. Archs Path. 93:356-365,1972.
9. Aschenbrenner, R., Bodechtel, G. Klin. Wschr. 17:298-302,1938.
10. Byer, E., Ashman, R., Toth, L.A. Am. Heart J. 33:796-806,1947.
11. Eichbaum, F.W, Gazetta, B.H., Bissetti, P.C., Pereira, C.B. Z. ges. exp. Med. 139:721-734,1965.
12. Greenhoot, J.H., Reichenbach, D.D. Neurosurgery 30:521-531,1969.
13. Weidler, D.J. Stroke 5:759-764,1974.
14. Graf, C.J., Rossi, N.P. J. Neurosurg. 49:862-868,1978.
15. O'Brien, M.D., Waltz, A.G. Stroke 4:201-206,1973.
16. Hall, R.E., Livingston, R.B., Bloor, C.M. J. Neurosurg. 46: 638-647, 1977.
17. Melville, K.I., Blum, G., Shister, H.E., Silver, M.D. Am. J. Cardiol. 12:781-791,1963.
18. de Molina, A.F., Hunsperger, R.W. J. Physiol. 160:200-213,1962.
19. Johansson, G., Olsson, K., Haggendal, J., Jonsson, L., Thoren-Tolling,K. Can. J. Comp. Med. 46:176-182,1982.
20. Guldin, W.O., Markowitsch, H.J. J. comp. Neurol. 229:393-418,1984.

ACKNOWLEDGEMENTS

This work was supported by the Heart and Stroke Foundation of Ontario. Dr. V. Hachinski is a Research Associate of the Heart and Stroke Foundation of Ontario and Dr. J. Ciriello is a Canadian Heart Foundation Scholar.

15

HEART RATE BEHAVIOUR DURING PARTIAL EPILEPTIC SEIZURES. AN ELECTROCLINICAL STUDY
ROSSI, E. (1) and ROSSI, G.F. (2)

Institutes of Cardiology (1) and of Neurosurgery (2), Catholic University, Rome, Italy

ABSTRACT

The behaviour of the heart rate during 75 partial epilepsy seizures was studied in adult patients. Seizure occurrence and character were analyzed by combining the recording of the electrocerebral activity (scalp EEG in all cases; direct brain recording through stereotactically implanted electrodes in 5 patients) and the TV monitoring of the clinical manifestations.
The heart rate changed significantly during the majority of seizures (61.3 %). Tachycardia was much more frequent than bradycardia.
The heart rate change occurred after the beginning of the seizure in 54,5 % of cases (mean delay 3,4 seconds); it was coincident with it in 9,1 %. No strict relation between cerebral site of origin of the ictal discharge and occurrence and sign of the heart rate change was found.
The possibility that a change of the heart rate is directly dependent upon the paroxysmal discharge of the epileptized cerebral neurons is suggested by the above findings.

INTRODUCTION

Visceral phenomena are a frequent component of the clinical manifestations of epileptic seizures. Changes in heart rate and rhythm have been reported by many (1-15). The cardiac functional involvement can reach such a degree as to cause death, at least in patients with previous - even if undetected - myocardial damage (16; see the review by Jay and Leestma, 17, reporting clinical and experimental findings). Little is known, however, on the nature of these phenomena. Are the cardiac functional variations directly produced by the abnormal cerebral discharge responsible for the seizure or are they indirectly related to it (for instance caused by the awareness of the seizure occurrence, or by the emotional content of the seizure, or by excessive motor activity)? Can both the sympathetic and the parasympathetic sections of the autonomic system become involved? In the case of partial epilepsy, is there any relationship between the anatomica- functional character of the epileptogenic cerebral zone and the occurrence and type of cardiac functional variation?

We thought that the above questions may be answered by an accurate study of patients suffering from partial epilepsy, by determining the relationship between the occurrence, time of onset and characteristics of the heart rate changes on one hand, and time of onset and site of origin of the ictal epileptic discharge and its resulting clinical manifestations on the other hand.

Detailed analysis of the electrocerebral epileptic abnormalities, and particulary of the ictal ones, combined with that of the clinical manifestations of the seizures provides the most reliable means to disclose the spatio-dynamic organisation of the epileptic process (18,19). However, the classical methodology recording the electroencephalogram by electrodes placed on the scalp, though very useful for revealing seizure occurrence, might not be sufficient to provide all the necessary informations in certain patients. In fact, the ictal discharges arising in the depth of the brain, such as the limbic structures, cannot be detected by scalp electrodes; the scalp electrodes would show the ictal activity only if and when it propagates from its deep site of origin to the more superficial brain regions or to the whole brain (secondary generalization) (Fig. 1). Consequently, one might get misleading information on both the time of onset and the site of origin of the ictal neuronal discharge. These difficulties might be to a great extent overcome by utilizing electrodes directly implanted within the brain, a diagnostic methodology largely used in neurosurgery in epileptic patients candidate to surgical treatment (18,19).

We report the preliminary results of research undertaken to try to answer the previously formulated questions.

MATERIAL AND METHODS

Seventy five epileptic seizures were analyzed. They were recorded in 14 adult patients (ranging in age from 15 to 43 years, mean 24.4 years) undergoing electrophysiological study in view of surgical treatment. All patients suffered from severe epilepsy of the partial type and in some cases with secondary generalization resistant to pharmacological treatment. The location of the cerebral site of origin of the neuronal discharge responsible for the seizure (the "lesional functional epileptogenic complex", 18) was inferred by combining the characters of the clinical seizure pattern with the topography of the electrocerebral abnormalities, and by taking into account all the available clinical and neuroradiological findings (see Fig. 7).

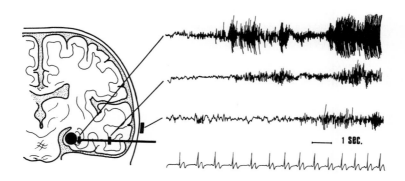

Fig. 1. Semischematic representation of a stereo-EEG recording from the temporal lobe. A seizure discharge (producing oral automatism), arises from the rhynoencephalic structures (first lead) and then propagates to the more superficial neurocortical temporal structures (second lead). A few seconds later it is detected from the scalp (EEG) electrodes (third lead). The heart rate changes (tachycardia) a few seconds after the beginning of the ictal discharge and more or less in conjunction with the appearance of the ictal discharge on the scalp EEG.

The nature of the brain damage responsible for epilepsy was variable (post-traumatic: 3 cases; ischemic: 1 case; post-encephalitic: 2 cases;

very slow developing tumor: 2 cases; unknown: 6 cases).

In all subjects the electrocerebral activity was recorded by electrodes conventionally applied on the scalp (EEG). In 5 of these, electrodes were also stereotactically and permanently implanted in preselected brain sites (stereo EEG or SEEG). In all patients thoracic electrodes were utilized to record the heart rate (ECG). Two 16-channel electroencephalographs were used for the EEG, SEEG and ECG recording (paper speed 15 mm/sec). The clinical manifestations of the epileptic seizure were TV monitored and recorded on TV tape. Each recording session lasted several hours and was repeated for several days in order to detect at least three seizures.

As far as the heart rate is concerned, two types of analysis were performed: 1) The first one was applied to the seizures monitored both with scalp EEG and with SEEG (75 seizures). Its first aim was to look for the occurrence of a change of heart rate as well as for its sign and degree; its second aim was to check whether occurrence and characters of heart rate variations were related to the cortical site of origin of the ictal discharge. The heart rate was estimated on the basis of 5-second-intervals. Three specimens of 30 seconds were considered: 2 minutes before, immediately before, and immediately after the beginning of the EEG seizure discharge (see Fig. 4). The second type of analysis was applied only to the seizures studied stereoelectroencephalographically (34 seizures). Its main purpose was to find out the precise time relation between the beginning of the electrocerebral ictal discharge and the beginning of the heart frequency change. The heart rate was measured by calculating the interval between successive beats (R to R), 15 seconds before and 15 after the appearance of the ictal discharge in the SEEG record (see Fig. 6). In both types of analysis the significance of the detected heart changes was evaluated using the Mann-Whitney test.

RESULTS

1) Occurrence and type of heart rate changes during partial epileptic seizures.

The rate of the heart during the seizure changed significantly (minimum significance $p < 0.05$) with respect to that recorded 2 minutes immediately before the seizure itself in all patients but one, and in the majority of the 75 seizures considered (46 out of 75).

Increase of heart rate (sinusal tachycardia) was by far the most frequent phenomenon (41 seizures, Figs. 1,2,5 and 7).

Bradycardia was rarely recorded (5 seizures, Figs. 3,5 and 7). More than a third of the seizures (29 cases) were not accompanied by significant heart rate variations (Fig. 7).

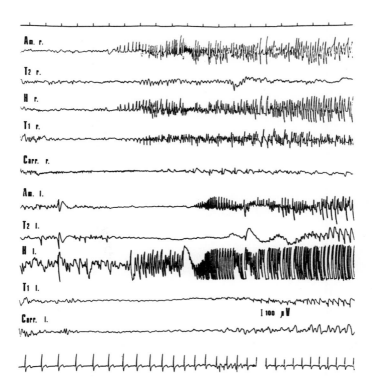

Fig. 2. Stereo-EEG recording of a seizure discharge which arises from the right amygdala (Am. r.) quickly propagates to the ipsilateral hippocampus (H r.) and temporal neocortex (T2 r., T1 r., Carr. r) and then to the contralateral hippocampus (H l.), amygdala (Am. l.) and temporal neocortex (T2 l., T1 l., Carr. l). The heart rate starts to change (tachycardia) at the very beginning of the seizure.

EEG MONOPOLAR

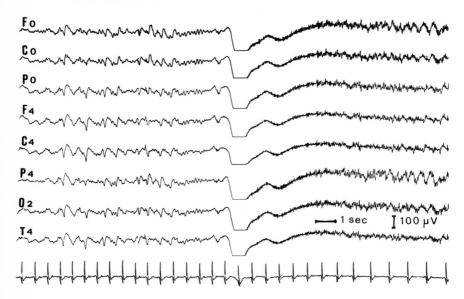

Fig. 3: Scalp EEG monopolar recording (reference on the neck) of
a seizure discharge from a large area of the right cerebral
hemisphere. The heart rate decreases (bradycardia) a few
seconds after the beginning of the ictal discharge.

The degree of the heart rate change observed was quite variable. Its
maximum was an increase of 37% with respect to the rate preceeding the
seizure (28% for bradycardia). The duration of the heart rate change was
similar to that of the seizure (Fig. 4, case G.E.). It rarely outlasted it
and would only occur if the seizure became secondarily generalized.
No other changes of the cardiac activity were observed besides the
above mentioned increase or decrease of the rate.

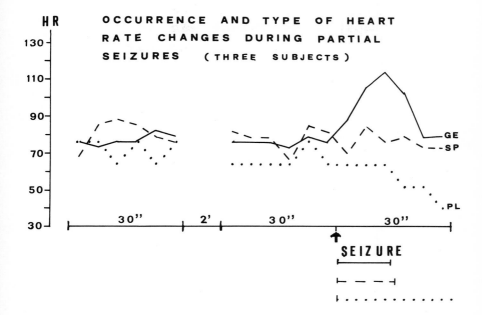

Fig. 4. Examples of the possible behaviour of heart rate (HR) during
partial seizures (three different cases). The onset of the
seizure is indicated by the arrow; its duration by the horizontal
line below it.

2) Time relation between onset of the ictal discharge and beginning of the
heart rate change.

This analysis was limited to the seizures studied
stereoelectroencephalographically and accompanied by significant heart rate
changes (22 seizures) in order to have the best opportunity to detect the
actual onset of the electrocerebral seizure discharge (see above). The
results are presented in Fig. 5.

In most cases (12 out of 22), the electrocerebral ictal discharge
started before any detectable variation of the heart rate (Figs. 1 and 3).
The mean delay was 3.4 seconds, the maximum delay was 10 seconds.

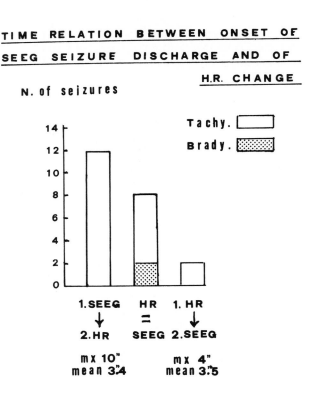

TIME RELATION BETWEEN ONSET OF
SEEG SEIZURE DISCHARGE AND OF
H.R. CHANGE

N. of seizures

Fig. 5: Time relationship between the onset of the electrocerebral
discharge reponsible for the seizure and the onset of the
heart rate change (in the twenty two seizures recorded with
stereo-EEG)

The beginnings of the cerebral and the cardiac events were considered
to be more or less coincident in 8 seizures (Fig. 2). Finally, the increase
of the heart rate was recorded before the electrocerebral and clinical
seizure onset in 2 instances (3 and 4 seconds respectively).
Exemplificative cases of the three possible time relationships between the
cerebral and cardiac phenomena are illustred in Fig. 6.

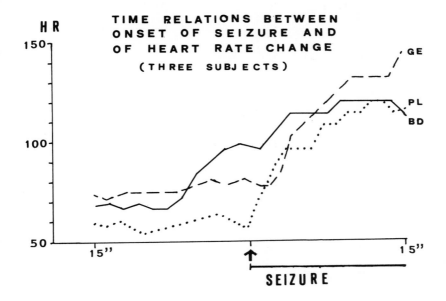

Fig. 6: Examples of the possible time relationship between the onset of
the ictal discharge responsible for the seizure (indicated by
the arrow) and the onset of the heart rate (HR) change.

3) Relationship between site of origin of the seizure discharge and heart
rate behaviour.

They were analyzed in all the 75 seizures. The results obtained are
schematically portrayed in Fig. 7. This shows the number of seizures as
well as the seizure - related heart rate behaviour for each cerebral region
harbouring the cerebral epileptogenic zone in our 14 patients.

Heart rate increase was found to accompany seizures originating from
all the affected brain sites with the exception of the motor area 4. It was
the only type of heart rate behaviour in the few seizures of occipital
origin (1 patient).

OCCURRENCE AND TYPE OF
HEART RATE CHANGES AND SITE OF
ORIGIN OF THE SEIZURES

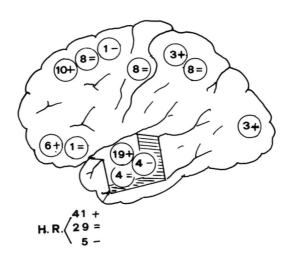

Fig. 7. Schematic representation of the location of the epileptogenic
zone (i.e. of the cerebral site of origin of the electrocerebral
discharge responsible for the partial seizure) in the 14 patients
considered. The numbers within the circles indicate the number
of seizure discharges recorded from the brain region (total: 75),
the signs +, = and - indicate the increase of heart rate (total:
41), no heart rate change (total: 29), and the decrease of heart
rate (total: 5) respectively.

Heart rate decrease was only recorded during 4 seizures of deep
temporal origin (3 patients) and 1 seizure from the frontal supplementary
motor areas (1 patient).

Finally, a relevant number of seizures arising from all the affected
brain sites, with the exception of the occipital lobe already mentioned,
were not associated with heart rate variations. Such a lack of heart rate
changes characterized the 8 seizures recorded from the motor area 4.

DISCUSSION

The occurrence of significant variations of cardiac function during epileptic seizures is confirmed by the present findings. We stress that our study was limited to the heart rate changes detectable during partial seizures, i.e. when the anatomo-functional abnormality responsible for epilepsy is confined to part of the brain.

In our material, the changes of cardiac activity which seemed related to the epileptic seizures were limited to the heart rate. We have to remark, however, that the technique we used to monitor cardiac activity was not appropriate for detailed electrocardiographic analysis.

As emphasized in the introduction, our main purpose was to try to get information on the nature of the variations of the cardiac function during the epileptic seizure. No conclusive evidence has been obtained. However, we believe that our findings permit some interesting remarks.

1) The following findings deserve particular consideration with relation to the origin of the seizure-related heart rate changes.
a) There was a close relationship between the duration of the electroencephalographic paroxysic discharge responsible for the clinical seizure manifestations and of the heart rate changes, provided that the former remained partial. The phenomenon was quite clear particularly when the seizure had a short duration (see Fig. 4).
b) The appearance of the heart rate change was coincident with that of the electrocerebral ictal activity in a consistent number of cases (4 out of the 5 patients studied with electrodes directly implanted within the brain, and 8 out of the 22 seizures which were accompanied by heart rate variations in these 4 patients).
c) In 5 seizures (3 patients) the seizure related variation of the cardiac activity was a slowig of heart rate.
d) Variations of the heart rate could accompany seizures with quite different clinical manifestations. On the other hand, the same clinical seizure pattern was associated with heart rate changes in some instances and not in others.

The four above-mentioned findings make it unlikely that the cardiac events considered were secondarily generated, i.e. provoked by some clinical manifestations of the seizure itself, as for instance, intense motor activity or psychic phenomena. They seem to indicate that in certain conditions the heart rate change has to be regarded as one of the clinical aspects of the seizure, namely a phenomenon directly depending on the paroxysmal discharge of the epileptized cerebral neurons.

The value of the suggested significance of these findings is based upon the assumption that the electroclinical approach to the study of our patients (see the methods) was apt to disclose the actual site of origin of the ictal discharges. Should our intracerebral electrodes have missed it, the recorded epileptic activity could have been propagated from elsewhere. In this case, the time of onset of the ictal discharge could have preceded the one shown in the recording. Actually, this might explain why in two instances the heart rate change seemed to precede the electrocerebral ictal discharge by a few seconds. However, on the basis of our experience of more than 100 SEEG explorations in epileptic patients, we believe that our approach provides reliable findings, if not in all, then certainly in the majority of cases (18,19).

The precedence of heart rate change appearance with respect to the electrocerebral (SEEG) and clinical signs of the seizures in the two cases mentioned above might have an alternative explanation. It is known that the epileptic attack can be precipitated by several factors of a different nature. In the two instances considered here, one might regard the precocious heart rate increase as a sign of some event (e.g. visceral or emotional) responsible for the triggering of the ictal cerebral discharge rather than as a consequence of it. A more complete and detailed study of the visceral activities during seizures would probably provide useful information.

Obviously, this does not rule out the possibility that at least some of the observed heart rate changes are only indirectly related to the cerebral epileptic discharges. Attempts were made to see whether the occurrence of heart rate changes is in some way related to the characters of the seizure clinical manifestation. This did not provide conclusive evidence. In any event, such a secondary origin should be taken into account, particularly when the cardiac rate starts to change several seconds after the beginning of the seizure.

2) If the above is accepted, then a second remark seems pertinent. The ictal discharge arising from the cerebral epileptogenic zone seems to

ultimately activate the sympathetic system (increase of heart rate) in the majority of cases. This view is in keeping with most of the reports of the literature (2,3,6,7,9-11,13,15,16). However, the heart rate was significantly reduced in 5 of the 75 examined seizures. In 2 of these 5 instances, the phenomenon was detected in patients studied stereoelectroencephalographically and its occurence was simultaneous with that of the electrocerebral ictal discharge. Therefore, an activation of the parasympathetic system initiated by the cortical neuronal ictal discharge seems possible as well (9-11).

3) We thus come to the last remark. We have seen that in the same patient seizures recorded at different moments can be accompanied by different heart rate changes or by insignificant variations. This variability of the autonomic response was already stressed (7,10). Our research shows that it can be found in spite of the site of origin of the related cerebral ictal discharge. This finding might indicate that the occurrence and the sign of the heart rate changes during seizures are not necessarily related to the location of the cerebral epileptogenic zone. Nevertheless (see Fig. 7); i) heart rate changes were never observed during the seizures arising from the motor area 4 (1 patient, 8 seizures); ii) 4 of the 5 seizures accompanied by bradycardia arose from the deep temporal structures. It is tempting to stress that the motor area 4 is a cortical region having a highly specialized somatic function and that the deep temporal structures are richly connected with the so called "visceral" (or limbic) brain. The literature on the variations of cardiac activity following direct electrical stimulation of the brain in man, although insufficient to provide complete information on this regard, clearly shows the relevant role of the limbic temporal and extratemporal structures (3,20-24).

More material would be necessary to validate the present findings. At the moment, the hypothesis suggests that the sign of the cardiac functional variations is dependent on several factors of which the location of the cerebral epileptogenic region is only one of them.

ACKNOWLEDGEMENT

This study was partially supported by the Ministry of Public Education.

REFERENCES

1. Gastaut, H. Epilepsia 2: 59-96, 1953.
2. Mulder, D.W., Daly, W., Bailey, A.A. Ann. Int. Med. 93: 481-493, 1954.
3. Penfield, W., Jasper, M. In: Epilepsy and the functional anatomy of the human brain. Little, Brown and Company, Boston, 1954.
4. Phizackerley, P.J.R., Poole, W.E., Whitty, C.W.M. Epilepsia 3: 89-91, 1954.
5. Mosier, J.M., White, P., Grant, P., Fisher, J.E., Taylor, R. Neurology 7: 204-210, 1957.
6. Metz. S.A., Halter, J.B., Porte, D., Robertson, R.P. Ann. Int. Med. 88: 189-193, 1978.
7. Van Buren, J.M., Ajmone Marsan C. Arch. Neurol. 3: 683-703, 1960.
8. White, P.T., Grant, P., Mosier J. Neurology 11: 354-361, 1961.
9. Bogacz, J., Yanicelli, E. Wld Neurology 3: 195-208, 1962.
10. Johnson, L.C., Davidoff, R.A. Electroencephalog. Clin. Neurophys. 17: 25-25, 1964.
11. Sulg, I.A. Electroenceph. Clin. Neurophysiol. 23: 389-396, 1967.
12. Mathew, N.J., Taori, G.W., Mathai, M.S., Chandy J. Neurology 20: 725-728, 1970.
13. Walsh, G.O., Masland, W., Goldensohn, E.S. Bull. Los Angeles Neurol. Sci. 37: 28-35, 1972.
14. Pritchett, E.L.C., NcNamara, J.O., Gallagher, J.J. Am. Heart J. 100: 683-688, 1980.
15. Marshall, D.W., Westmoreland B.F., Sharbrough F.W. Mayo Clin. Proc. 58: 443-446, 1983.
16. Leestma, J.E., Kalelkar, M.B., Teas S.S., Jay, W., Hughes, J.R. Epilepsia 25 (1): 84-88. 1984.
17. Jay, G.W., Leestma, J.E. Acta neurol. scand., suppl. 82, 63: 1-66, 1981.
18. Rossi, G.F., Gentilomo, A., Colicchio, G. Schwartz. Arch. Neurol. Neurochir. Psychiat. 115: 229-270, 1974.
19. Rossi, G.F., Colicchio, G., Gentilomo, A., Pola, P., Scerrati, M. Arch. Ital. Biol. 120: 160-175, 1982.
20. Pool, J.L., Ransohoff, J. J. Neurophysiol. 12: 385-392, 1949.

21. Delgado, J.M., Mihailovich, L., Sevillano, M. J. Neurol. Ment. Dis. <u>130</u>: 477-487, 1960.

22. Heath, R.B., Mickle, W.A. <u>In</u>: Electrical studies of the unaesthetized brain, (Eds. E.R. Ramey, D.S. O'Dogherty), Hoebur, New York, pp. 214-217, 1960.

23. Roberts, L. <u>In</u>: Electrical stimulation of the brain (Ed. D.G. Sheer), Univ. Texas Press, pp. 533,553, 1961.

24. Van Buren, J.M. J. Neurosurg. <u>18</u>: 273-288, 1961.

PART FOUR

CARDIOEMBOLIC STROKE

5

16

CARDIOEMBOLIC STROKE: CONCEPTS AND CONTROVERSIES

HART, R. G.

Department of Medicine, (Neurology), University of Texas Health Science Center, 7703 Floyd Curl Drive, San Antonio, Texas, U.S.A. 78284

The concept of thrombotic fragments detaching from the heart and migrating to occlude distal arteries has been recognized for more than a century. In 1828, Allibert repcrted blockage of a femoral artery by a coagulum similar to one found in the left atrial appendage (1). In 1847, Virchow described occlusion of cerebral arteries by thrombi that appeared to arise in the heart (2). Virchow later termed this phenomena embolism, derived from the Greek word for plug. In 1852, Kirkes reported a patient whose middle cerebral artery was plugged by a "firm, oval mass about the size and shape of a grain of wheat... the mitral valve was much diseased, the auricular surface beset with large warty excrescenses" (3).

The first case series of cardioembolic stroke in which the clinical and pathological features were critically analyzed appears to be that of John Lidell in 1873 (4). In this remarkable report, Lidell noted that the symptoms "almost always suddenly develop" and that these emboli "generally lodge in the middle cerebral artery." Neuropathologically, he reported that "it is very striking that in cases of red softening, the foci are habitually in relation with an embolism." Lidell proposed that preventive measures be employed in those patients who have heart diseases which predisposed to brain embolism, suggesting citrate of ammonia to "diminish the coagulability of the blood" (4).

From these auspicious beginnings, clinical concepts about cardioembolic brain ischemia have advanced slowly in the ensuing century, but at an ever-increasing rate. It was not long ago that cardiogenic emboli were considered an uncommon cause of ischemic stroke. Technologic progress has allowed reliable, noninvasive identification of many potential cardiac sources of emboli during life. It is now appreciated that about one of every six brain infarcts is due to this mechanism (5).

Stober, et al. (eds.), CENTRAL NERVOUS SYSTEM CONTROL OF THE HEART. Copyright © 1986 by Martinus Nijhoff Publishing. All rights reserved.

As many potential cardiac sources are identifiable prior to stroke, stroke prevention does indeed seem possible, as proposed by Lidell over 100 years ago. However, two major problems limit preventive efforts. Firstly, only a fraction of patients with cardiac disorders prediposing to embolism will ultimately experience stroke. Unless subgroups of patients at particularly high risk of embolism can be identified, many patients with potentially embolic cardiac disorders may need to be treated to prevent stroke that is destined to occur in only a minority. Secondly, many of the currently available preventive treatments carry a substantial morbidity (e.g. anticoagulation). Thus, in attemping to prevent cardioembolic stroke, it must be certain that the treatment is not worse than the disease. Consequently, despite burgeoning clinical data defining the natural history of cardioembolic stroke, effective preventive measures have been largely lacking.

Clinical studies are limited by the inability to diagnose cardioembolic stroke with certainty in individual patients, despite the increasing capacity to image safely the heart and the cerebral vasculature. There is no practical, clinical "gold standard" for the clinical diagnosis of cardioembolic stroke. Many patients with ischemic stroke have both a potential cardiac source of emboli and concomitant cerebrovascular disease. It is difficult, and at times impossible, to determine which mechanism accounts for stroke in individual patients. While specific clinical features (e.g. abrupt onset of maximal neurologic deficit, absence of ipsilateral TIA) may favor cardiogenic embolism, quantitative correlation of clinical features with autopsy confirmation is not available. Even autopsy findings can be inconclusive, due to disappearance of embolic fragments. The lack of quantitatively valid clinical criteria for cardioembolic stroke is a major problem in research efforts and in individual patient management.

There is little doubt that nonrheumatic, nonvalvular atrial fibrillation is a marker of increased risk of ischemic stroke. Several studies, the most convincing from the Framingham investigators, have shown that people with atrial fibrillation have about five times the risk of stroke as controls, matched for age, sex and hypertension (6). This increased stroke risk approaches that which follows TIA. However, it is uncertain whether these strokes are due to cardiogenic embolism or to coexisting cerebrovascular disease. Perhaps atrial fibrillation is only a marker of other cardiac disease or cerebrovascular disease that are direct cause of brain ischemia. The mechanism(s) of brain ischemia in patients with atrial fibrillation is an important and unsettled clinical dilemma, with clear implications regarding stroke prevention.

In the past 18 months, several clinical studies have continued to forge the links between acute myocardial infarct, ventricular dyskinesis, mural thrombus formation, and embolic stroke. While ischemic stroke follows in only about 3% of unselected patients with acute myocardial infarct, the stroke risk may reach 25% within 3 months of myocardial infarcts complicated by mural thrombi (5, 7, 8). Further studies of stroke prevention in this setting are needed. In contrast, thrombi identified in ventricular aneurysms remote from acute myocardial infarct appear to have a relatively low embolic potential. In patients who experience brain ischemia associated with ventricular mural thrombi remote from myocardial infarct, careful consideration should be given to other mechanisms before attributing stroke to cardiogenic embolism.

The association between mitral valve prolapse and stroke has been further explored by several investigators (5, 9). The Framingham investigators performed M-mode echocardiography on 4967 people, and detected mitral valve prolapse in 5% of their population (10). While about 4% of men of all ages had mitral valve prolapse, women under age 30 had an 18% prevalence (10). In another report, there was discouragingly high interobserver variability when three experienced cardiologists interpreted the same echocardiograms for the presence or absence of mitral valve prolapse (11). This variability emphasizes the importance of blinded interpretation of echocardiograms in clinical studies purporting to relate stroke to mitral valve prolapse.

About 20% of patients with native valve infective endocarditis experience brain embolism (5, 12, 13). Patients with valvular vegetations demonstrated echocardiographically may be at particular risk (14). It is uncertain if anticoagulation is of overall benefit in patients with native valve endocarditis, and most clinicians do not routinely anticoagulate such patients at present. However, patients with mechanical prosthetic valve endocarditis have a stroke risk approaching 50% if not anticoagulated (5). Anticoagulation appears to reduce this risk, and careful continuation of anticoagulation may be indicated in patients with mechanical prosthetic valve endocarditis (5, 12).

The risk vs benefit of early anticoagulation of patients with acute cardioembolic stroke is controversial. Early anticoagulation following acute embolic stroke can potentiate hemorrhagic transformation (15). The optimal time for initiation of anticoagulation and risk factors for hemorrhagic transformation will be further explored at this Workshop. Alternative forms of antithrombotic therapy (i.e., low molecular weight heparinoids, tissue plasminogen activator, newer platelet antiaggregation agents) have not been studied in cardioembolic stroke. Similarly, the optimal intensity of chronic oral anticoagulation for prevention of embolic stroke is unknown. The current therapeutic range is largely based on anecdote or extrapolated from venous system thromboembolism. The optimal type and intensity of antithrombotic therapy for the prevention of cardiac thrombi may vary depending upon the particular cardioembolic source.

So despite being over a century old, clinical concepts of cardioembolic stroke remain rudimentary. Little has been proven and controversies abound. Retrospective, observational data are subject to methodologic criticism and are often unconvincing. Yet such observations are the conceptual groundwork defining important areas for further study. The many problems and uncertainties surrounding cardioembolic stroke, which at times seem overwhelming, are the challenges that make clinical research in this area particularly exciting.

REFERENCES

1. Allibert, P.C. Recherches sur une occlusion peu connue des Vaisseaux arteriels consideree comme cause de gangrene. Paris, 1828.
2. Virchow. Arch. Path. Anat. p.272, 1847.
3. Kirkes, W.S. Transactions of the Medico-Chirurgical Society (London), pp. 281-324, 1852.
4. Lidell, J.A. Treatise on Apoplexy. William Wood and Co., New York, pp. 169-214, 1873.
5. Cerebral Embolism Task Force. Cardiogenic brain embolism. Arch. Neurol. 43: 71-84, 1986.
6. Wolf, P.A., Dawber, T.R., Thomas, H.E. et al. Neurology 28: 973-977, 1978.
7. Weinreich, D.J., Burke, J.F., Pauletto, F.J. et al. Ann. intern. Med. 100: 789-794, 1984.

8. Keating, E.C., Gross, S.A., Schlamowitz, R.A. et al. Am. J. Med. 74: 989-995, 1983.

9. Jackson, A.C., Boughner, D.R., Barnett, H.J.M. Neurology 34: 784-787, 1984.

10. Savage, D.D., Garrison, R.J., Devereaux, R.B. et al. Am. Heart J. 106: 571-576, 1983.

11. Wann, L.S., Gross, C.M., Wakefield, R.J. et al. Am. Heart J. 109: 803-808, 1985.

12. Carpenter, J.L., McAllister, C.K. Sth. med. J. 76: 1372-1375, 1983.

13. Foster, J., Hart, R.G. Neurologic complications of bacterial endocarditis: A reappraisal. Neurology (submitted).

14. O'Brien, J.T., Geiser, E.A. Am. Heart J. 108: 386-394, 1984.

15. Cerebral Embolism Study Group. Stroke 15: 779-789, 1984.

17

DOES CARDIOEMBOLIC STROKE HAVE A NEUROLOGIC PROFILE?

RAMIREZ-LASSEPAS, M., CIPOLLE, R.J., BJORK, R.J., KOWITZ, J.J., WEBER, J.C., STEIN, S.D., SNYDER, B.D.

University of Minnesota; Dept of Neurology; St. Paul-Ramsey Medical Center; 640 Jackson St; St. Paul, MN 55101

ABSTRACT

The clinical symptoms at onset and neurological findings of 193 consecutive patients with an acute cerebral infarct were analyzed to determine if the frequency of their occurrence allowed one to distinguish among patients with a cardiac source of embolus (SOE) (106 patients), an aterial SOE (38 patients) or no demonstrable SOE (49 patients). Rapidity of onset and loss of consciousness at onset were the only symptoms significantly more frequent in the group of patients with a cardiac SOE.

INTRODUCTION

The diagnosis of cardioembolic stroke is made on the basis of the identification of a potential cardiac source of embolus (SOE) and the occurrence of two or more neurologic symptoms from a group that, through the years, has become the "clinical criteria" for diagnosis of cerebral embolism. These criteria are:
1. sudden onset of neurologic deficit
2. occurrence of neurologic deficit during activity
3. restricted focal deficit (i.e. isolated aphasia, monoplegia)
4. loss of consciousness (LOC) at onset
5. headache as a prominent symptom at onset
6. convulsive seizure or seizures at onset
7. nausea and/or vomiting at onset
8. peak of neurologic deficit at onset
9. evidence of two or more neurologic deficits of simultaneous onset or close in time of onset
10. evidence of systemic emboli (kidney, intestine, limb)

To test the validity of the use of these symptoms as criteria for diagnosis and determine if indeed cardioembolic stroke presents with a distinct neurologic profile, we conducted the following investigation.

METHODS

The clinical symptoms at onset and neurologic findings on examination of 193 consecutive patients hospitalized after acute onset of focal neurologic deficits and diagnosed as having had an acute cerebral infarct (ACI) were correlated by repeated cross tabulations to the presence of a cardiac SOE, an arterial SOE or the absence of a demonstrable SOE. The following findings were considered as a potential cardiac SOE: (1) supraventricular arrhythmias (atrial fibrillation with or without valvulopathy or enlarged left atrium, atrial flutter, supraventricular tachycardia, wandering pacemaker, brady-tachycardiac arrhythmias, sick sinus syndrome), (2) ventricular PVC's, ventricular tachycardiac, (3) myocardial infarction (acute, subacute with arrhythmia, old with mural thrombus, ventricular aneurysm or arrhythmia), (4) other (atrial myxoma, endocardial tumors, prosthetic valves, endocarditis, prolpasing mitral valve). Ascending aortic aneurysm, aortic dissection, atherosclerotic ulcerations of the aortic arch and atherosclerotic ulceration and/or stenosis of the carotid and/or vertebral arteries were considered potential arterial SOE.

RESULTS

There were 49 patients in whom no SOE could be identified (Group A); in 38 (Group B), an arterial SOE was identified; and, in 106 (Group C), a potential cardiac SOE was found - in 22 of these, a potential arterial SOE was also identified. Table 1 shows the cardiac SOE identified in Group C and its frequency.

Repeated cross tabulations were performed by computer to compare occurrence of symptoms at onset and its combination with each of the patient groups. The only associations found to have statistical significance were the rapidity of onset and LOC at onset, both with group C (Table 2). The occurrence of sudden LOC at onset was highly specific for Group C (11 of 12 such events) but not sensitive. (Only 10.3% of Group C patients experienced such an onset). Patients with a cardiac SOE were, on the average, older and had a larger number of intercurrent hospital events and higher mortality than patients in Groups A and B.

From the above findings, we conclude that there is no reliable, predictable clinical pattern of presentation of ischemic stroke in patients with a possible cardiac SOE and that possibly, except for sudden LOC as a presenting symptom, there is no clinical neurologic profile for cardioembolic stroke.

Table 1

CARDIAC SOURCE OF EMBOLUS

Cardiac Abnormality	Number
Atrial Fibrillation (ASHD* old MI)	45
Atrial Fibrillation (RHD**)	8
Atrial Fibrillation (AMI***)	8
Mulitple Arrhythmias (AMI)	8
Ventricular Arrhythmias (AMI)	6
Brady-Tachy-Arrhythmias	6
Wandering Pacemaker Flutter	4
Flutter-Fibrillation	4
Multifocal PVC's	3
Anteroseptal AMI	3
Prosthetic Valve	3
Mural Thrombus, Old MI	2
Congenital Heart Disease	2
Bigeminy	2
Ventricular Aneurysm	1
Marantic Endocarditis	1

   * Atherosclerotic heart disease (ASHD)
  ** Rheumatic heart disease (RHD)
*** Acute myocardial infarction (AMI)

Table 2

ASSOCIATION BETWEEN RAPID ONSET, LOSS OF CONSCIOUSNESS AT ONSET,
AND PRESENCE OF A SOURCE OF EMBOLUS

	Rapid Onset	%	LOC	%	Both	%
GROUP A	30	62.5	0	0.0	0	0.0
GROUP B	20	55.5	1	3.0	1	3.0
GROUP C	81	80.2*	16	19.3**	11	13.2***
	131	70.8	17	10.2	12	7.2

  * p<0.03
 ** p<0.001
*** p<0.001

18

THE COMMON COINCIDENCE OF CAROTID AND CARDIAC LESIONS

HACHINSKI, V.C., REM, J.A., BOUGHNER, D.R. and BARNETT, H.J.M.

Departments of Clinical Neurological Sciences and Cardiology, University of Western Ontario, London, Ontario, Canada

ABSTRACT

All of 184 consecutive patients with TIA (68) and cerebral infarction (116) admitted to an investigative stroke unit had cardiac monitoring with a Hewlett Packard 78524 arrhythmia monitoring system. In addition, 127 underwent 2-D echo cardiography and 113 had conventional or digital intravenous cerebral angiography. From the clinical history and examination and all the cardiac investigations, 59 patients (32 %) (21 with TIA's and 38 with strokes) had a possible cardiac source of emboli. In 30 (16 %) (9 TIA and 21 stroke patients) the heart was the most likely source of cerebral emboli but after cerebral angiography 29 of the 59 patients (16 %) also showed a vascular lesion in the appropriate carotid artery so that no definite decision could be made regarding the source of emboli. Cardiac and carotid lesions with embolic potential often coexist in the same patient, making a full investigation mandatory before an etiology can be ascribed. Even then, an open mind and close follow up are required to assure appropriate management.

INTRODUCTION

We had been struck by the common coincidence of cardiac and carotid lesions in patients with cerebrovascular diseases. In order to analyze this problem systematically, we studied two hundred and nineteen consecutive patients admitted from January 1 to December 31, 1983, to the Investigative Stroke Unit, University Hospital, London, Canada, with the diagnosis of TIA and cerebral infarction. Thirty-five patients (16 %) with other diagnoses were excluded so that 184 patients were the subject of this study.

METHODS

The history was taken and the general and neurological examinations were carried out by a neurology resident (speciality trainee). There were 182 patients (98,9 %) who had an admission electrocardiogram (ECG). All patients were monitored for a minimum of 48 hours with a Hewlett Packard 7525 Arrhythmia Monitoring System. Fifty-five patients (29.9 %) also had additional 24 to 48 hour Holter monitoring using a two Channel Recorded by Zymed, because of the suspicion of an arrhythmia. One hundred and twenty-seven patients (69 %) underwent 2-D echocardiography with a Hewlett Packard 77020A Ultrasound Imaging System. One hundred and seventy-six patients (95.7 %) had computerized tomography (CT) of the head (General Electric 8800 CT scanner). One hundred and thirteen patients (61.4 %) underwent cerebral angiography of which 67 patients had the procedure performed by a transfemoral catheter technique (angiography) and 57 patients by digital intravenous angiography (DIVA). Eleven patients had both investigations. The patients' profile for age, sex, TIA and stroke is shown in table 1.

Table 1

Age and Sex of the Patients

	Men	Women	Total
N	122 (66.3%)	62 (33.7%)	184
TIA	47	21	68 (37%)
Stroke	75	41	116 (63%)
Mean Age	62.8	64.8	63.5
Range	25-86	30-83	25-86

RESULTS

One hundred and eighty-two patients (98.9 %) had ECG on admission. In 108 patients, the ECG was abnormal (59.3 %) and in 74 patients (40.7 %) normal. Thirteen patients had AF, ten of them known previously. Forty-two patients had signs of myocardial infarct (MI) on ECG. Nineteen patients had a history of a recent or remote MI. Twenty-three had a silent MI. Thirty-one of 42 patients with MI (73.8 %) were non-diabetics (NDM) and eleven patients (26.2 %) suffered from diabetes mellitus (DM). Patients with DM had significantly more MI's than NDM (Chi-square test, p<0.05). By comparing only the silent MI, the significance for having an MI in the DM-group becomes even higher (Chi-square test, p<0.01). The mean age in NDM-group with MI on the ECG was 67.0 years (± SD 9.6), range 40-81 and 63.3 years (± SD 8.2), range 48-77 in the DM-group. The patients in the DM-group with MI on ECG were significantly younger (t.test, p<0.05). Fourteen non-diabetics with a silent MI (9.1 %) had a mean age of 68.1 years (± SD 7.8), range 48-76 (10 patients 65 years) and nine diabetics with a silent MI (30 %) had a mean age of 61.1 years (± SD 7.9), range 52-77 (2 patients 65 years). Diabetics with a silent MI were significantly younger than non-diabetics (t.test, p<0.05).

The 48-hour cardiac monitoring was abnormal in 94 patients (51.1 %) and normal in 90 patients (48.9 %). Of the seventeen patients with AF, eleven were persistent and six paroxysmal. Eleven of these (eight persistent, three paroxysmal) were known by history, two patients with persistent AF were detected by ECG and confirmed by the cardiac monitoring, and in four patients the abnormalities were detected by the monitoring. Two patients had a 2 heart block type Mobitz II. One of them had a normal ECG and the other had a 10 AV-block. One patient with a transient 3 heart block had a normal resting ECG.

The 2-D echocardiography was abnormal in 61 patients (48 %), normal in 56 patients (44.1 %) and ten investigations (7.9 %) were difficult to evaluate because of technical problems.

The mean age of the 61 patients was 65.3 years (± SD 11.2), range 30-83. There are only two patients younger than 45 years, both in the MVP group. The following eight patients were known to have abnormalities by history or previous 2-D echocardiography: MVP (2), prosthetic mitral (2) and aortic (1) valves, aortic valve sclerosis (2) and segmental ventricular disease (hypokinetic segment) (1). A previously unknown possible cardiac source for emboli was detected in 22 patients (17.3 %), table 2, mean age 64.7 years (± SD 9.8), range 43-81, only one of these patients was less than 45 years old (in the MVP group).

One hundred and seventy-six patients had a CT of the head (95.7 %). In 98 patients (55.7 %), the CT was abnormal and in 78 patients (44.3 %) normal. In 80 patients (81.6 %) of which 71 had stroke and 9 had TIA, the CT findings were related to the clinical presentation and in 18 patients

(18.4 %) of which 11 had TIA and seven had stroke, they were not related.

Table 2

Echocardiography and Possible Cardiac Source for Emboli

11 Segmental ventricular
    disease   - 2 with thrombus
                                    - 1 with aneurysm and thrombus
                                    - 1 mitral annulus calcification
 7 Mitral valve prolapse   - 1 with vegetation (endocarditis)
                                    - 1 mitral annulus calcification
 2 Aneurysm   - 2 with thrombus
 1 Thrombus   - with CAD known because of previous
                                    echocardiogram

 1 Global myocardial
    dysfunction
22 Patients (17.3 %)

One hundred and thirteen patients (61.4 %) had cerebral angiography. The investigations were abnormal in 90 patients (78.6 %) and normal in 23 patients (21.4 %). The angiographic lesions were related to the clinical presentation in 73 patients (81.1 %) and unrelated in seventeen patients (18.9 %). Sixty-seven patients had cerebral angiography done by the transfemoral catheter technique (59.3 %) and 57 patients (50.4 %) DIVA. Eleven patients had both investigations, two of them with normal results. Stenosis from mild (0-30 % narrowing) to very severe (90 % narrowing) was present in 48 patients (42.5 %), occlusion of an artery in fourteen patients (12.4 %) and atherosclerotic changes without stenosis in eleven patients (9.7 %). All these lesions were in vessels appropriate to the symptoms and signs. There were 29 patients who had a lesion in the appropriate carotid and they also had a possible cardiac source for embolus.

DISCUSSION

Franco et al. (1) demonstrated that patients with cerebrovascular accident and TIA frequently have echocardiographic abnormalities (58 %), many unsuspected clinically. They suggested that echocardiography should be performed in these patients since the cardiac abnormalities identified may be contributory to the cerebrovascular event.

The incidence of unsuspected echocardiographic abnormalities in an older adult population remains unclear. Most echocardiographic studies exclude cardiovascular disease when studying an aging population, focusing on alterations in ventricular function and chamber size. A study carried out in our city with participation by one of the authors (D.R.B.) did list the incidence of abnormalities on M-mode echocardiography in the elderly (2). Among 146 asymptomatic volunteers with a mean age of 72 years (range 60-94 years), 38 had findings on history and physical examinaion that excluded them from the echocardiographic study. Of this excluded group, 10 (6.8 %) were hypertensive, two (1.4 %) were diabetic, six (4.1 %) had left ventricular hypertrophy on their electrocardiogram, nine (6.1 %) had previous myocardial infarcts and two had evidence of aortic stenosis. An additional nine were excluded because of obstructive lung disease, hyperthyroidism and intermittent claudication. Of the remaining 108 patients who underwent M-mode echocardiography, 14 (13 %) produced unsatisfactory studies compared with the 7.9 % failure rate in our two dimensional echocardiographic study. The M-mode studies showed nine patients (9.5 %) to have unsuspected mitral valve prolapse compared with 7.6 % in our series. No examples of mitral stenosis were found and three patients showed mitral annular calcification. One structurally abnormal aortic valve was noted. Also in that study, there were no instances of unsuspected septal or posterior wall motion abnormalities compatible with previous infarction. However, the ability of M-mode echocardiography to detect coronary artery disease is limited since it only images the left ventricular minor axis and does not examine either the apex or the antero-lateral wall. Conceivably, silent myocardial infarction in those two areas may have been missed. Our present study showed a much higher incidence of unsuspected wall motion abnormalities in the TIA patients as well as various valvular lesions. Only the mitral valve prolapse cases were expected on the basis of the results from the healthy elderly volunteers.

From the clinical history and all cardiac investigations we found 59 patients (32 %) with a possible cardiac source for cerebral emboli. After cerebral angioraphy, 20 of those 59 patients were found also to have a vascular lesion in the appropriate carotid and we could not decide definitely which lesion really was responsible for the cerebral embolus. In the remaining 30 patients (15.4 %), we believe that the heart was the most likely source.

Our results show that cardiac investigations are worthwhile in TIA and stroke patients, who are considered candidates for preventive treatment. Only by evaluating both the carotid and cardiac status can one make an informed judgement as to the likely cause of a TIA or cerebral infarct in individual patients.

REFERENCES

1. Franco, R., Alam, M., Ausman, J. et al. Circulation 62 (Suppl III): 22, 1980.
2. Manyari, D., Patterson, C., Johnson, D. et al. J. Clin. Exp. Gerontology 4 (4): 504-520, 1982.

ACKNOWLEDGEMENTS

This study was supported by grants from the Heart and Stroke Foundation of Ontario. Drs Boughner and Hachinski are Research Associates of the Heart and Stroke Foundation of Ontario.

19

SEQUENTIAL CHANGES OF REGIONAL CEREBRAL BLOOD FLOW IN EMBOLIC AND
THROMBOTIC CEREBRAL ARTERY OCCLUSION

MINEMATSU, K., YAMAGUCHI, T., CHOKI, J. and TASHIRO, M.

Cerebrovascular Division, Department of Internal Medicine, National
Cardiovascular Center, 5-7-1 Fujishirodai, Suita, Osaka 565, Japan

ABSTRACT

This study was designed to investigte differences of sequential
changes of rCBF between embolic and thrombotic cerebral artery occlusion.
Sixty eight cases with embolic (embolic group) and 63 cases with thrombotic
cerebral artery occlusion (thrombotic group) were analyzed using Xe-133
inhalation technique.
In embolic group, mean rCBF of the affected hemisphere was significantly
lower than that of the unaffected hemisphere only in subacute (Day15-Day28)
and chronic stage (Day29-). Futhermore, mean rCBF of the affected
hemisphere in acute stage (Day1-Day14) was significantly higher than that
in chronic stage. In thrombotic group, mean rCBF of the affected hemisphere
was significantly lower than that of the unaffected hemisphere, and was
fairly constant in all stages.
When mean rCBF of each patient in acute stage was plotted against that in
chronic stage, regression lines of thrombotic group were approximately
equal to the line of Y(acute)=X(chronic) in both sides, while mean rCBF of
embolic group in acute stage was higher than that in chronic stage, making
regression line Y=0.47X+23.1 in the affected hemisphere.
High rCBF in acute stage of embolic group of the present study was thought
to be reflection of an increase of blood flow after reopening of the
previously occluded artery, that is frequently observed in cerebral
embolism. Thus, sequential changes of rCBF in cases with cerebral embolism
appeared to be quite different from those with cerebral thrombosis.

INTRODUCTION

Although our knowledge concerning cerebral blood flow and metabolism in ischemic cerebrovascular diseases has markedly increased, it is not easy to understand the cerebral circulatory and metabolic states in each patient with cerebral infarction. As mentioned in our previous report, for example, there are several differences in the pathophysiological and neuroradiological features between embolic and thrombotic cerebral infarction (1). Therefore, it is reasonable to infer that there are some differences of cerebral circulatory and metabolic states between these two subtypes of cerebral infarction.

The purpose of the present study is to clarify the following questions. Firstly, is there any difference in the sequential changes of regional cerebral blood flow (rCBF) between embolic and thrombotic cerebral artery occlusion? Secondly, if there is a difference, what kinds of significance does it have in the pathophysiology of the cerebral infarction?

SUBJECTS AND METHODS

Subjects of the present study were patients with unilateral supratentorial cerebral infarction with embolic or thrombotic occlusion of a cerebral artery. The diagnosis of cerebral embolism was made when the patient met at least two of the following criteria, (1) abrupt onset of focal neurological symptoms and signs, (2) existence of a source of emboli, (3) evidence of embolization to other parts of the body. Atrial fibrillation is so common in elderly patients that it was not considered as an independent criterion unless visualization of an embolus and/or reopening of previously occluded vessel were confirmed by cerebral angiography. Patients not meeting these criteria but having arteriographic evidence of occlusion of cerebral arteries were diagnosed as cerebral thrombosis. Equivocal and undetermined cases were excluded from this study. Presumed cerebral thrombosis in the territory of perforating arteries was also excluded if occlusion of cerebral artery was not visualized by angiography.

Regional cerebral blood flow of these patients was measured by Xe-133 inhalation technique at various times after onset. Patients were categorized by timing of measurements into three groups; an acute stage group measured before day 14, a subacute stage group, measured between day 15 to day 28, and a chronic stage group measured on or after day 29. In more than a half of these patients, rCBF measurements were repeated at least twice at different stages.

The number of cases, age, size of infarct and other parameters of each group are summarized in Table 1. Infarct-Index in this table is a ratio of the largest hypodense area to the hemispheric area that was computed on CT films to represent a size of the infarct. There were no significant differences in age, Infarct-Index and PaCO2 between embolic and thrombotic group in each stage.

RESULTS

The time course of cerebral blood flow measurements in both groups is shown in Fig. 1. In the thrombotic group, mean hemispheric rCBF (mCBF) of the affected hemisphere was significantly lower than that of the unaffected hemisphere and fairly constant in all stages. In contrast, mCBF of the affected hemisphere in the embolic group during the acute stage was not lower than that of the unaffected side and was significantly higher than that in the chronic stage.

When mCBF of each case in the acute stage was plotted against that in the chronic stage, regression lines of both the affected and the unaffected hemisphere in thrombotic group were approximately equal to the line of Y=X, while mCBF in the acute stage of embolic group was higher than that in the chronic stge, making the regression line Y=0.47X+23.1 in the affected hemisphere (Fig. 2).

Patterns of rCBF distribution were classified into three subtypes, that is, global ischemia, focal ischemia and focal hyperemia. In 5 of 12 embolic cases focal hyperemia in the acute stage turned into focal or global ischemia in the chronic stage. In the thrombotic group only one case showed focal hyperemia in the acute stage and the majority showed a focal or global ischemic pattern from the acute stage.

DISCUSSION

In the present study the most distinctive feature of rCBF in cerebral embolism was an increase in blood flow during the acute stage and with cerebral thrombosis was a decrease of cerebral blood flow in all stages.

Using Xe-133 intracarotid injection method, it has been reported that

	Embolism			Thrombosis		
	Acute	Subacute	Chronic	Acute	Subacute	Chronic
No. of cases	32	25	45	18	22	45
Age (y.o.)	58.7±12.7	59.0±11.4	55.5±15.5	62.1± 9.4	62.4± 8.4	60.8±11.4
Infarct-Index (%)	16.9±15.7	20.8±16.1	20.0±16.0	14.6±15.2	13.2±12.3	12.7±13.0
Sites of occlusion						
ICA	3	7	9	3	9	17
MCA-stem	6	4	8	12	8	19
MCA-branch	15	10	15	2	1	4
Miscellaneous	8	4	13	1	4	5
Day of measurement						
mean	8.3	22.3	76.7	6.4	21.6	110.8
range	2-14	15-28	29-543	1-12	15-28	29-945
$PaCO_2$ (mmHg)	38.0± 4.3	39.0± 4.0	39.2± 3.3	38.7± 3.7	38.5± 3.2	39.5±3.2
MABP (mmHg)	89.9±15.6	91.8±12.8	89.5±11.4	103.2±20.2	103.0±12.1	100.5±19.0

(mean ± S.D.)

Table 1. Summary of subjects by the timing of rCBF measurement.

Time Course of Cerebral Blood Flow

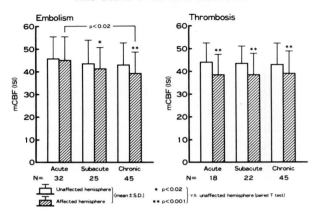

Fig. 1. Mean hemisperhic rCBF (mCBF) of the affected and the unaffected
hemisphere and their sequential changes. Values of mCBF are
represented by the initial slope index (ISI). In the embolic group,
mCBF of the affected hemisphere (44.5 $\pm$ 10.3) is not significantly
different from that of the unaffected hemisphere (45.3 $\pm$ 9.6) in
the acute stage, but higher than that in the chronic stage (39.5 $\pm$
9.3, $p < 0.02$).

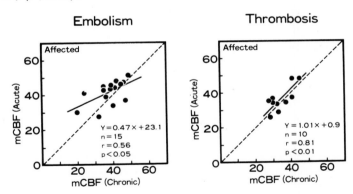

Fig. 2. Relationships between mCBF in the acute stage (ordinate) and that
in the chronic stage (abscissae). In the majority of cases with
cerebral embolism, mCBF in the acute stage is higher than that in
the chronic stage, suggesting the presence of hyperemia in the
acute stage of cerebral embolism.

focal hyperemia in the acute phase of cerebral infarction is frequently observed in cases without arterial occlusion or with reopening of the previously occluded artery (2-4). A "capillary blush" on angiogram is observed in some cases of cerebral infarction with reopening of the occluded artery, and is thought to be a reflection of postischemic hyperemia (5). Thus, it should be remembered that reopening of the occluded artery may play one of the important roles in the pathogenesis of focal hyperemia in the acute stage of cerebral infarction.

Complete disappearance of arteriographic occlusion of cerebral artery or migration of emboli to the distal part of the artery is a characteristic phenomenon frequently observed in cerebral embolism. In our 120 consecutive cases with cerebral embolism, 28 of 42 cases who had repeated angiography showed complete reopening. When cases with incomplete reopening (migration of emboli) were included, reperfusion of the ischemic zones occurred in 90 % of the cases (1).

Increased CBF of the affected hemisphere in the acute stage of cerebral embolism is thought to be "post-ischemic" hyperemia following reopening of the occluded artery. This concept is supported by the recent report by Skyhøj Olsen and Lassen (6), in which they suggested that the migration of the embolus is associated with partial or complete reperfusion leading to hyperemia in the initially ischemic brain tissue. Although information about cerebral metabolism was lacking in these studies, focal hyperemia in cerebral embolism might partly have prompted the development of the concept of "the luxury-perfusion syndrome" by Lassen (7).

SUMMARY

Sequential changes of rCBF in cases with cerebral embolism are somewhat different from those in cerebral thrombosis. Cases with cerebral embolism have increased rCBF of the affected hemisphere in the acute stage, while those with cerebral thrombosis have a fairly constant and pronounced decrease of rCBF of the affected hemisphere in all stages. The former is thought to be a reflection of an increase of CBF after reopening of the previously occluded artery.

REFERENCES

1. Yamaguchi, T., Minematsu, K., Choki, J. et al. Jpn. Circ. J. 48:
   50-58,1984.
2. Paulson, O.B., Lassen, N.A., Skinhøj, E. Neurology (Minneap) 20:
   125-130,1970.
3. Uemura, K., Goto, K., Ishii, K. et al. Neuroradiology 16:228-232,1978.
4. Skyhøj Olsen, T., Larsen, B., Skriver, E.B. et al. Stroke 12:598-607
   1981.
5. Irino, T., Taneda, M., Minami, T. Neurology (Minneap) 27:471-475,1977.
6. Skyhøj Olsen, T., Lassen, N.A. Stroke 15:458-468,1984.
7. Lassen, N.A. Lancet ii:1113-1115,1966

20

TIMING OF HEMORRHAGIC TRANSFORMATION OF CARDIOEMBOLIC STROKE

CEREBRAL EMBOLISM STUDY GROUP

HART, R.G.

Department of Medicine (Neurology),University of Texas Health Science Center, 7703 Floyd Curl Drive, San Antonio, Texas, U.S.A. 78284

ABSTRACT

Sixty-two patients with cardiogenic brain embolism and secondary hemorrhagic transformation by CT were retrospectively studied to determine the timing of hemorrhagic transformation. Thirty-four patients were receiving anticoagulants and 28 were not. In nonanticoagulated patients, hemorrhagic transformation usually occurred on CT done more than 6-18 hours after stroke. No patient with hemorrhagic transformation was known to have a nonhemorrhagic CT at 48 hours.
Hemorrhagic transformation is usually delayed for several hours after embolic stroke, but is usually present by 48 hours.

It has long been known that cardioembolic brain infarcts are expecially prone to hemorrhagic transformation (1). Hemorrhagic transformation of initially pale, embolic infarcts occurs both in patients who are receiving anticoagulants and those who are not (2, 3). Initiation of anticoagulation therapy is often delayed in patients with acute cardioembolic stroke, for fear of potentiating secondary brain hemorrhage.

We retrospectively collected 62 patients with aseptic cardioembolic stroke and secondary brain hemorrhage, either hemorrhagic infarct or hematoma, to define the clinical features of hemorrhagic transformation. The diagnosis of cardioembolic brain infarct required a defined cardiac source of emboli, a stroke syndrome in which embolism was deemed the likely cause, and an initial computed tomographic (CT) scan showing infarct, either hemorrhagic or nonhemorrhagic. In patients who subsequently developed hematoma, an initial, nonhemorrhagic CT was required to document the secondary nature of the hemorrhage. Infarct size was estimated based on clinical features at the time of initial evaluation (3).

Thirty-four patients were receiving anticoagulants at the time of hemorrhagic transformation, and 28 patients were not. Clinical deterioration associated with hemorrhagic transformation occurred in 15 (44%) of 34 anticoagulated patients compared to 2 (7%) of 28 patients who were not anticoagulated. The timing of spontaneous hemorrhagic transformation in these 28 patients who were not receiving anticoagulation is considered in detail. Clinical predictors of hemorrhagic worsening in the 17 patients who deteriorated during hemorrhagic transformation is considered in an accompanying abstract.

Of these 28 patients who experienced hemorrhagic transformation in the absence of anticoagulation, only two experienced recognized clinical worsening (indicated on Figure 1 at time "X"). In the remainder, the time of hemorrhagic transformation must be given as an interval, with extremes defined by stroke onset and/or CT data. Patients in whom the interval extends to time zero had hemorrhagic infarct on their initial CT (Figure 1). In these patients, hemorrhagic transformation must have occurred between stroke onset and initial CT. The remaining patients had an initial, nonhemorrhagic CT followed by a second CT showing hemorrhagic infarct or hematoma. The infarct size was estimated as large in 68%, moderate in 25%, and small in 7%. Infarct size had no apparent relationship to the timing of hemorrhagic transformation.

Most of these infarcts had undergone transformation within 48 hours of stroke onset. The graph in Figure 2 is generated from the time intervals in Figure 1. The area indicated as "?" reflects the interval between nonhemorrhagic and hemorrhagic CTs, in which the exact time of onset of hemorrhagic transformation is uncertrain. For example, 12 hours after stroke onset, at least 10% and possibly as many as 84% of these infarcts had undergone hemorrhagic transformation. By 48 hours after stroke onset, at least 68% and perhaps as many as 100% of infarcts had hemorrhagically transformed.

By another method of analysis of the data shown in Figure 1, it is appreciated that none of the 7 CTs done between 3-6 hours after stroke detected hemorrhagic infarction. Of the 11 CTs done between 6-18 hours after stroke, 45% detected hemorrhagic infarction. Thus, the CTs carrie

onset (n=18) will usually not identify infarcts that are destined to undergo spontaneous hemorrhagic transformation (sensitivity = 28%). Of CTs carried-out between 24 and 48 hours after stroke onset, the sensitivity was 87%.(Fig. 1.)

Figure 1.

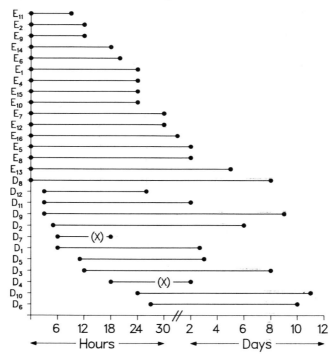

Timing of Hemorrhagic Transformation
— Nonanticoagulated Patients —

As a methodologic caveat, these data largely reflect hemorrhagic transformation occurring during the initial 10-14 days following cardioembolic stroke. Asymptomatic hemorrhagic transformation that may have occurred later, following discharge from an acute care hospital ward, may be systematically under-represented in these data. Nevertheless, it is hemorrhagic transformation within these initial days which may be the critical determinant of safe initiation of anticoagulation. We conclude that most hemorrhagic transformation in cardioembolic strokes occurs within

48 hours of stroke onset, but that very early CT (18 hours from stroke onset) will not reliably identify infarcts that are destined to become spontaneously hemorrhagic.

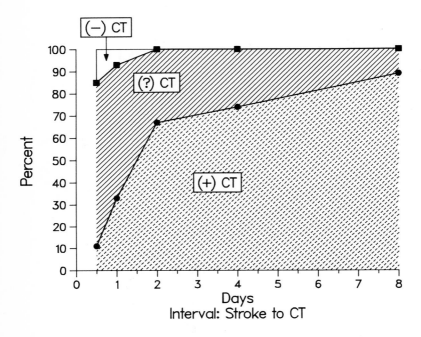

Figure 2. Likelihood of detection of hemorrhagic infarction by CT. Data derived from Figure 1; see text for explanation.

REFERENCES

1. Fisher, C.M., Adams, R.D. J. Neuropath. exp. Neurol. 10: 92-93, 1951.
2. Cerebral Embolism Study Group. Stroke 15: 779-789, 1984.
3. Cerebral Embolism Study Group. Stroke 14: 668-676, 1983.

21

PATHOMECHANISM OF CARDIOEMBOLIC STROKE

HORNIG, C.R.

Department of Neurology, Justus Liebig University, Giessen, FRG

One of the most interesting questions concerning the pathogenesis of embolic stroke is, what is the fate of the embolus and what are the sequeli? Pathoanatomic and angiographic studies have provided evidence for some possibilities of clinical importance.

If an embolus remains at its site of occlusion for some time, antegrade and retrograde thrombotic material may occur. Alterations of blood viscosity, as observed by Dr. Coull (1) in patients with cardioembolic stroke may be of importance. This phenomenon has especially been observed in cases of embolism into the intracranial portion of the internal carotid artery. Spread of thrombosis into the origins of the anterior and middle cerebral arteries can cut off collateral circulation of the affected hemisphere and greatly increase the area of infarction.

Several courses may be taken by an embolus resulting in restoration of blood flow in a portion of the ischemic area. The first is migration of the embolus. The reason for migration up the affected vessel might be vasodilatation caused by acidosis in the hypoxic area. After the embolus has passed, recirculation of the former ischemic region may occur. Migrating emboli have been observed in serial angiograms and Fisher and Adams (2) saw hemorrhagic transformation predominantly in cases with evidence for migration of the embolus in their pathologic study.

The second mechanism for restoration of blood flow in the ischemic area is fragmentation of the embolus. Some angiographic studies have shown that embolic occlusion of the trunk of the middle cerebral artery has changed to multiple occlusions of its branches a few days later.

The third way is lysis of the embolus or its fragments. Enzymes of white blood cells may play a role in this process as well as release of tissue plasminogen activator from the vessel wall.

A further possibility to account for the recirculation of infarcted tissue, not restricted to the embolic type of stroke, is by collateral pathways which may come into play with diminishing edema.

Restoration of blood flow in embolic stroke generally occurs early in the course of the disease, sometimes within hours, predominantly within the first days. The exact peak frequency isn't known as yet, but perhaps the use of transcranial doppler sonography may provide further information.

Restoration of blood flow in cases of embolic stroke becomes apparent by angiographic findings or blood flow measurement as reported by Dr. Minematsu (3).

Furthermore, ischemic damage of the blood brain barrier becomes more evident in cases of temporary ischemia as in many cases of embolic stroke. This explains the well known observation of more frequent and earlier positive isotope scans, contrast enhancement by CT or raised protein concentration in the cerebrospinal fluid in embolic stroke.

A clinically more important sequel of recirculation is hemorrhagic transformation of the infarct (Table 1 and 2). Mostly petechial hemorrhages are observed, seldom intracerebral hematoma. The latter might be the result of large confluent petechial hemorrhages or of the rupture of a larger vessel damaged by the ischemia. We have evidence for both possibilities both clinically and by CT.

Table 1

Frequency of Hemorrhagic Transformation

Study Design	Number Studies	Number Patients	Hemorrhagic Transformation	Range
Pathoanatomic	2	511	24 %	18-42 %
CT (retrospective)	12	2680	7 %	3-37 %
CT (prospective)	6	241	21 %	4-43 %

Table 1 contains the results of some pathoanatomic and CT studies regarding the frequency of hemorrhagic infarction. The wide range is accounted for by the fact that most studies didn't consider either the etiology of the ischemic event or the timing of CT scanning. Controversies exist regarding the predominant time of transformation. The results of Hart and Sherman (4) are in accordance with ours showing transformation in cardioembolic stroke within the first few days after the ischemic event.

Table 2

Risks for Hemorrhagic Transformation

	Animal Experiments	Pathologic Studies	CT and Clinical
Large infarcts	+	+	+
Arterial chronic	+		
hypertension acute	+	+	
Venous pressure	+	+	
Anticoagulants	+	+	+/-

Table 2 shows the risk for hemorrhagic transformation found in different studies. In a prospective study we were able to show an increased risk for hemorrhagic transformation in cases of large infarcts, especially with a mass effect, severe neurological deficit and disturbance of consciousness, in cases of involvement of the cortex, and contrast enhancement by CT and blood-CSF barrier disturbance.

REFERENCES

1. Coull, B.M., Beamer, N.B., Seaman, G.V.F. This sxmposium. Abstract J. Neurol. 3: 197, 1985.
2. Fisher, C.M., Adams, R.D. J. Neuropath. exp. Neurol. 10: 92-94, 1951.
3. Minematsu et al. This volume.
4. Hart, R.G., Sherman, D.G. This volume.

22

RECURRENT CEREBRAL EMBOLISM AND FACTORS RELATED TO EARLY RECURRENCE
- ANALYSIS OF 186 CONSECUTIVE CASES

YAMAGUCHI, T., MINEMATSU, K., CHOKI, J., MIYASHITA, T. and OMAE, T.

Cerebrovascular Division, National Cardiovascular Center, 5-7-1 Fujishirodai, Suita, Osaka 565, Japan

ABSTRACT

Clinical profiles of cerebral embolism were studied in relation to the factors and incidence of recurrent attacks in a series of 186 consecutive cases, who were diagnosed as certain cerebral embolism based on our diagnostic criteria.

Atrial fibrillation (AF) without valvular heart disease (59 cases) and valvular heart disease (58 cases) were the two main underlying cardiac disorders.

More than 90 % of the remaining 69 cases were found to have various types of cardiac diseases.

Within 2 weeks prior to the cerebral episode which caused admission to our hospital, 24 patients (13 %) had experienced 27 episodes of cerebral (14) and/or systemic (13) embolization, suggesting an importance of minor episodes as a warning sign of major strokes. During acute stage (-2 wks) recurrent cerebral episodes were observed in 0.51 %/patient-day and systemic embolization in 0.81 %/pt-day, which were significantly more frequent than those in subacute (2-4 wks) and chronic (4-8 wks) stages.

Fourteen percent of recurrent cerebral and 18 % of systemic embolization were attributable to direct cause of death.

Incidence of early recurrent episodes to the brain in patients with AF was approximately the same as that in VHD patients, but early systemic embolization occurred more frequently in VHD than in AF patients.

Early anticoagulation appeared to be effective as far as the recurrent episodes were concerned. On the contrary administration of urokinase in acute stage seemed to increase the recurrence rate.

INTRODUCTION

The secondary prevention of ischemic cerebrovascular disease is a subject of considerable importance as is primary prevention since recurrent attacks markedly increase the functional disability and mortality of stroke patients (1, 2). To better manage these patients, it is necessary to clarify the precise clinical profile including that of recurrent attacks, of each subtype of cerebral infarction; i.e., lacunar stroke due to perforating artery thrombosis, cerebral infarction due to thrombotic occlusion of major arteries and embolic occlusion, especially of cardiac origin. To this end the incidence, timing and prognosis of recurrent attacks and those predisposing factors relating to recurrence in cerebral embolism of cardiac origin were studied in a large series of cerebral infarctions.

SUBJECTS AND METHODS

The subjects of the present study were 186 consecutive patients with cerebral embolism diagnosed by our criteria (3) and admitted to our institution over a period of 6 years. More than 95 % of these cases were admitted within one week after onset.

The diagnosis of cerebral embolism was made when the patient met at least two of the following criteria: (1) abrupt onset of maximal focal deficit, (2) presence of a potential source of emboli, including (i) valvular heart disease (VHD), cardiomyopathy, acute myocardial infarction, infective endocarditis, etc.., ii) idiopathic or nonvalvular atrial fibrillation (NVAF), sick sinus syndrome, etc.., (3) evidence of embolization to other parts of the body. Since many of the elderly patients with atrial fibrillation have arterioslcerotic vascular disease, atrial fibrillation was not taken as an independent item unless the presence of an embolus or reopening of a previously occluded vessel were confirmed by cerebral angiography.

There were 95 men and 91 women, their ages ranged from 14 to 86 years, with a mean age of 61.4 years. Underlying cardiac disorders included 59 with NVAF, and 58 with VHD. The remaining 69 patients had various embolic sources such as cardiomyopathy, acute myocardial infarction, infective endocarditis, etc.

The incidence of recurrent cerebral and systemic embolization were expressed as percent per patient-day in which the total number of episodes was divided by the total number of days observed during the following four periods; (1) 2 weeks prior to the episode prompting admission, (2) day 1 - day 14 (acute stage), (3) day 15 - day 28 (subacute stage), and (4) day 29 - day 60 (chronic stage). The diagnosis of an embolic episode prior to admission was made only if the patient had a reliable history of focal neurological deficits or abdominal pain, hematuria or melena of sudden onset. After admission a sudden focal neurologic deterioration associated with changes of CT and possibly angiographic findings was considered as evidence of a recurrent brain embolus. Renal scintigraphy and renal or abdominal angiography results were taken into account for the diagnosis of systemic embolization in addition to ischemic symptoms and signs of each organ.

RESULTS

1. Preceeding episodes

During a two week period preceeding the episode causing the patients' hospitalization, 24 patients (13 %) experienced 27 episodes of cerebral and/or systemic embolism. Among them, 14 episodes in 14 patients were cerebral, and 13 episodes in 11 patients were systemic emboli. These results suggest the importance of minor episode as a warning sign of a major stroke in patients who have a potential cardiac source of emboli.

2. Recurrent attacks after admission

In the acute stage, recurrent cerebral episodes occurred in 0.51 %/patient(pt)-day, and systemic embolism in 0.81 %/pt-day. In the subacute and chronic stages, the frequencies of recurrent episodes to the brain were 0.24 and 0.12 %/pt-day, and those of recurrent systemic embolism were 0.05 and 0.04 %/pt-day, respectively. The incidences of recurrent cerebral and systemic embolization in the acute stage were significantly higher than those in the chronic stage.

The patient recurrence rate within two weeks (acute stage) was 14 % for both cerebral and systemic embolism, and 7 % for cerebral embolism alone. The reported incidence of recurrent cerebral and systemic embolization during the same period varies from 2 to 20 %, with an average of around 15 % (4-7). Recurrence rates limited to the brain in the literature are also variable probably because of the small number of cases in each reported series (8-10).

In this series more than 90 % of cerebral and 40 % of systemic recurrent emboli caused a deterioration in the patient's condition including 3 deaths in each group (23 % and 16 %, respectively). Only one fatal case was seen among the patients with recurrent attacks during the subacute and chronic stage.

Thus, it is apparent that in the acute stage of cerebral embolism, the probability of a recurrent embolus is much higher than in the subacute and chronic stages, and that recurrent embolization during this stage has tremendous deliterious effects on outcome.

3. Factors related to early recurrence (Fig. 1)

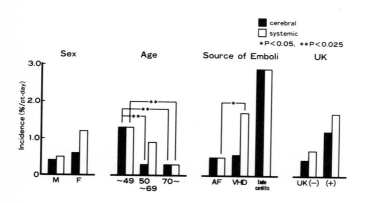

Figure 1. Factors related to early recurrence of cerebral embolism.
        Incidences are expressed as %/pt-day.
        UK: Urokinase

The incidence of early recurrence (within 2 weeks) tended to be higher in women than in men for both cerebral and systemic embolism, but the differences were not significant.

When the patients were divided into 3 age groups of less than 49, 50 to 69 and over 70 years old, the incidence of early recurrence to the brain in the youngest group (1.3 %/pt-day) was significantly higher than that of the two older age groups (0.3 %/pt-day in each group). For systemic embolism in the acute stage, the youngest group also showed a higher incidence (1.3 %/pt-day) than the oldest group (0.3 %/pt-day).

Early recurrence rates in VHD and AF were approximately the same for cerebral embolization (0.57 and 0.50 %/pt-day, respectively), but those for systemic embolism were significantly higher in VHD (1.70 %/pt-day) than in AF (0,50 %/pt-day). When cerebral and systemic recurrences were combined, VHD tended to have a higher rate than AF. However this difference did not reach a statistically significant level. Although infective endocarditis showed a high recurrence rate, it was not taken into account because of the small number of patients.

Urokinase was used in 17 patients during the acute stage. Early recurrence rates for cerebral and systemic embolism in these cases were approximately twice as high as those in cases without urokinase. Statistical analysis was not performed because of marked differences in the numbers of patients with and without urokinase.

Thus, age and type of embolic source appeared to be possible factors related to early recurrent embolism. However, the proportion of patients with VHD, AF and other cardiac disorders was somewhat different in each age group, therefore differences in early recurrence rate by age group might be merely a reflection of the type of embolic source. Using the hypothesis that the recurrence rate in each type of embolic source was not influenced by age, an expected recurrence rate was computed in each age group using three major cardiac disorders (VHD, AF and endocarditis) as follows:

$$\text{Expected recurrence rate} = \frac{\sum_{i=1}^{n} X_i Y_i}{\sum_{i=1}^{n} Y_i}$$

where, X_i = recurrence rate in patients with disorder "i"
Y_i = total number of days of observation in patients with disorder "i"

Although both the actual and expected recurrence rates of cerebral and systemic embolism were highest in the youngest age group, the actual recurrence rate of this group was higher than that expected suggesting that age could also be one of the factors promoting recurrent embolism early after the initial embolic stroke (Table 1).

Age Group	Cerebral		Systemic	
	expected	actual	expected	actual
- 49	1.0	2.2	1.9	3.1
50 - 69	0.6	0.4	1.2	1.2
70 -	0.5	0.2	0.6	0.6

Table 1. Expected and actual recurrence rate by age groups in acute stage (day 1-14).
Incidences are expressed as %/pt-day.

SUMMARY

The results of the present study are summarized as follows:
(1) Nearly 15 % of patients with cerebral embolism had experienced minor cerebral and/or systemic embolic episodes during the 2 weeks prior to the major embolic stroke prompting admission to our institution.
(2) Recurrent attacks after the major stroke occurred most frequently within 2 weeks.
(3) Age of the patients (younger age) and type of embolic source (VHD) appeared to have a close relationship to early recurrent attacks of cardiogenic embolism.

To clarify the factors related to early recurrent attacks and to establish the preventive measures for it, further prospective studies using a larger series will be necessary.

REFERENCES

1. Daly, R., Mattingly, T.W., Holt, C.L. et al. Am. Heart J. 42: 556-581, 1951.

2. Carter, A.B. Lancet 2: 514-519, 1965.

3. Yamaguchi, T., Minematsu, K., Choki, J. et al. Jpn. Circ. J. 48: 50-59, 1984.

4. Darling, R.C., Austin, W.C., Linton, R.R. Surgery Gynec. Obstet. 124:106-114, 1967.

5. Hart, R.G., Coull, B.M., Hart, D. Stroke 14: 688-693, 1983.

6. Calandre, L., Ortego, J.F, Berbemo, F. et al. Arch. Neurol. 41: 1152-1154, 1984.

7. Sherman, D.G., Goldman, L., Whiting, R.B. et al. Arch. Neurol. 41: 708-710, 1984.

8. Koller, R.L. Neurology 32: 283-285, 1982.

9. Sage, J.I., Van Uitert, R.L. Stroke 14: 537-540, 1983.

10. Cerebral Embolism Study Group. Stroke 14: 668-676, 1983.

23

A PROSPECTIVE STUDY ON THE RISK OF IMMEDIATE ANTICOAGULATION IN CARDIAC EMBOLIC STROKE

LODDER, J.

Dept. of Neurology, Medical Faculty of the University of Limburg, Maastricht, The Netherlands

ABSTRACT

Ninety-five patients with cardiogenic brain embolus underwent either early (n=69) or delayed (n=26) anticoagulation (AC). There were two recurrent emboli both in the early AC group. There were no cerebral bleeding complications despite the presence of large cerebral infarcts in 25 of the early AC group patients.

INTRODUCTION

The timing of anticoagulant (AC) treatment after a cardiac embolic stroke remains a matter of dispute. It is not clear whether a beneficial effect of early AC treatment outweights the risk of cerebral bleeding. Since we were not convinced of a substantial risk of cerebral bleeding complications, in March 1981 we started a prospective study to investigate the risk of immediate AC treatment in patients with a cardioembolic stroke.

METHODS

Included in the study were patients with a non-septic supra tentorial cardiac embolic stroke, lasting longer than 24 hours. Patients received AC treatment within 3 weeks following the cerebral event. Prior to AC a CT was done. Cardiac embolic stroke was diagnosed in the presence of 1 or more of the following features: transient or persistent atrial fibrillation, rheumatic heart disease, aortic or mitral valve stenosis or prosthesis, myocardial infarction in the preceeding 2 weeks, or left ventricular aneurysm. All patients were seen by a cardiologist at least once. Regarding CT and clinical findings, a distinction was made between 'small deep infarcts' and 'cortical infarcts'. Cortical infarcts were divided in small, moderate or large infarcts. Immediate full AC consisted of a standard heparin treatment, i.e., an intravenous bolus of 5,000 i.u. followed by a daily dose of 20,000 i.u. heparin by constant i.v. infusion until oral AC (OAC) was therapeutic (TT: 5-15%). When full AC was begun beyond 24 hours after stroke onset, treatment was regarded as delayed. Exclusion criteria for AC were general contraindications such as high blood pressure, bleeding tendency, hepatic failure. Furthermore, impaired consciousness and hemorrhagic infarction (HI) on CT were excluded. From April 1982 patients with infarcts with space occupying effects indicated by a shift of midline structures on CT were also excluded from initial AC. Any neurological deterioration was investigated by CT to look for bleeding complications. Follow-up was 3 weeks after stroke onset.

RESULTS

There were 69 patients with full AC within 24 hours (group I). Fifty-seven received heparin; 12 were already on AC and had a thrombo test percentage within the therapeutic range. In 26 patients (group II) treatment was delayed for various reasons: i.e., elevated blood pressure, decreased consciousness, hemorrhagic infarction on CT, or because paroxysmal atrial fibrillation was detected subsequently. There were 17 lacunar strokes in group I and 6 in group II, while the number of cortical strokes was 52 and 20 respectively. The estimated infarct size of cortical strokes is shown in Table 1. In 10 group I patients treatment was stopped after a median of 4 days for various reasons: 1 patient developed hemorrhagic infarction on repeated CT without clinical worsening, 1 had poor mental state. Three patients developed impaired consciousness while the CT did not show hemorrhagic complications in two. Two of these patients died some days later. The third patient had a large hemorrhagic infarction on repeated CT. One patient had hepatic dysfunction and 3 more patients died. In one patient treatment was stopped without any specific reason. Thirty-five repeated CT scans were made during full AC after a median of 5 days after stroke onset. Three patients showed HI, in two without clinical worsening, while the third patient had a gradual increase in neurological deficit, probably caused by edema. During the 3 weeks of follow-up period 5 patients died: one from recurrent cerebral embolism and one from a pulmonary embolus. One patient died from pneumonia and one from intestinal hemorrhage. In this last patient the pulmonologist continued antiocoagulant treatment because of a suspicion of pulmonary embolus despite the fact that the attending neurologist strongly advised against the treatment because the blood pressure was too high and the patient had decreased consciousness. Recently we had a second death caused by intestinal hemorrhage. The last patient had recurrent embolism to the brain but most probably died of bilateral pneumonia confirmed at autopsy. In this patient there were multiple hemorrhagic infarctions at autopsy. The other 4 patients had no hemorrhagic signs at autopsy. So none of these patients died because of hemorrhagic cerebral complications. Table 2 summarizes the effects of AC on recurrence rate and the side effects. There were no recurrent emboli and no complications of AC treatment in the delayed group (II).

DISCUSSION

In this prospective study on the risk of early AC treatment in 69 patients with a cardiac embolic stroke, there were no cerebral bleeding complications, not even in the 25 patients with large infarcts. Large infarcts are considered to be at especially increased risk for hemorrhagic transformation and therefore bleeding complications are most prevalent in patients with large infarcts. Therefore, as a general rule, we agree with others that in patients with large infarcts anticoagulant treatment should be postponed for at least some days. In our 23 AC treated patients with small deep infarcts, no cerebral bleeding complications occurred. Since a number of small deep infarcts visible on CT might be caused by . a cardiac embolus, we suggest that in the presence of a cardiac embolic source such patients should be anticoagulated early, especially in the absence of hypertension and when the CT scan reveals such infarcts. In the 69 early AC patients there were 2 recurrences (3 %). The estimated recurrence rate in the first two weeks is 10 to 20%. Therefore, our data suggest that early AC treatment lowers the chance of early recurrences. Our data shows that in the absence of general contraindications, no HI on CT, no mass effect on CT, and if the patient is alert, early AC treatment can be safely started. In patients with large infarcts, AC treatment should be postponed for at least some days since HI and/or cerebral edema will develop in a number of patients with a risk of cerebral bleeding, and most of these patients cannot be identified in advance.

Table 1:

cortical strokes	total	estimated infarct size		
		small	moderate	large
group I	52	7	20	25
group II	20	2	6	12

Table 2:

	no of patients	recurrent embolism	neurol. worsening due to AC	clinical worsening due to AC
early AC	69	2	0	1*
delayed AC	26	0	0	0

* death from intestinal hemorrhage

24

BRAIN HEMORRHAGE IN EMBOLIC STROKE

SHERMAN, D.G., HART, R.G. for the CEREBRAL EMBOLISM STUDY GROUP

The University of Texas Health Science Center at San Antonio, Department of Medicine (Neurology)

ABSTRACT

Brain hemorrhage is a feared complication of immediate anticoagulation of cardiogenic brain embolism. The clinical features of 62 patients with CT evidence of brain hemorrhage complicating embolic stroke were studied. Worsening with hemmorhagic transformation occurred in 15 anticoagulated and two non-anticoagulated patients, with no apparent relationship with patient age or embolic source. Of anticoagulated (AC) patients who worsened, 7 were hypertensive, 11 were excessively anticoagulated and 9 had large infarcts. There were 3 AC patients who worsened who were neither hypertensive, nor excessively anticoagulated. Asymptomatic hemorrhagic transformation was more common than hemorrhagic worsening, occurring in 45 patients. Factors associated with hemorrhage and worsening include large infarct size (60 % vs 21 %) and initiation of AC after early (less than 12 hrs from stroke onset) CT (62 % vs 37 %). The occurrence of symptomatic brain hemorrhage in embolic stroke is not entirely predicable. In patients with large embolic strokes, postponing anticoagulation for several days may be prudent.

INTRODUCTION

It is now generally accepted that patients having suffered a cerebral or systemic embolism from a cardiac source should be anticoagulated. The as yet unresolved issue is the timing of initiation of anticoagulation. The goal is to anticoagulate the patient with a cardiembolic stroke adequately enough to prevent recurrent emboli but to avoid the promotion of brain hemorrhage. The aim of this study was to analyze the development of brain hemorrhage in patients with cardioembolic stroke in hopes of discovering features that might predict a group at high risk for this complication.

METHODS

Subjects of Study

The subjects of this study were patients with aseptic cardioembolic brain infarcts who had an initial CT scan free of hemorrhage and subsequently developed hemorrhagic transformation confirmed by CT scan. The criteria for cardioembolic stroke were: (1) sudden onset of a maximal focal neurologic deficit (2) absence of evidence of cerebral vessel diseases, e.g. no antecedent TIA, no bruits, and when available no evidence of extracranial vascular disease by angiography or noninvasive studies, and (3) the presence of a cardiac disorder capable of producing intercardiac thrombus. The cases were gathered retrospectively from the case records of the members of the Cerebral Embolism Study Group from 14 medical centers. A data base was generated from the record of each patient with particular attention to factors considered to be potentially relevant in the genesis of brain hemorrhage. Some of these factors included age, sex, presence of hypertension, nature of the cardiac disorder, size of the brain infarction, and the nature and timing of anticoagulation.

RESULTS

Sixty-two patients were identified who met the criteria of this study. All patients developed hemorrhagic infarction (HI) or brain hematoma after an initial CT scan free of hemorrhage. All patients had a presumed cardiac source for their clinical embolic stroke. Forty-five (73 %) of these patients had no clinical worsening associated with the development of hemorrhagic infarction by CT scan. Seventeen (27 %) of the 62 patients had clinical worsening in conjunction with the CT development of brain hemorrhage. These seventeen patients with clinical worsening are the subjects of the present study. The mean age of these patients was 60 years (range 29-83); there were 10 men and 7 women. The underlying cardiac disorder was nonvalvular atrial fibrillation in 8, rheumatic valvular disease in 5, myocardial infarction in 2, mitral valve prolapse in 1 and prosthetic valve in 1. Fifteen (88 %) of the 17 patients with worsening were receiving anticoagulants. The 15 patients who worsened while receiving anticoagulants were compared to the 24 patients anticoagulated as a part of the randomized study of anticoagulation performed by the Cerebral Embolism

Study Group (Table 1). Large cerebral infarctions occurred in 9 (60 %) of the 15 patients who developed brain hemorrhage with clinical worsening whereas only 5 (21 %) of the 24 patients without hemorrhage or worsening had large infarcts. A larger portion (60 %) of the patients with brain hemorrhage were anticoagulated on the basis of a CT scan without evidence of hemorrhage obtained early, i.e., within 12 hours, following their embolic stroke. In contrast only 37 % of patients with an uneventful course following anticoagulation had their initial CT scan before 12 hours. Excessive anticoagulation was a poor predictor of risk of brain hemorrhage with worsening. A partial thromboplastin time of greater than three times control was noted in 67 % of the patients who worsened and in 77 % of those who did not. There were no deaths in the group without hemorrhage whereas 47 % of those with hemorrhage and worsening died.

Table 1   BRAIN HEMORRHAGE IN CARDIOEMBOLIC STROKE

	Anticoagulated With Worsening	Anticoagulated No Hemorrhage
Number	15	24
BP >180/100	7 (47 %)	--
Large infarct	9 (60 %)	5 (21 %)
Initial CT Interval	18 hrs	20 hrs
% < 12 hrs	60 %	37 %
Anticoagulation		
Initiated < 24 hrs	46 %	42 %
PTT > 3 x control	67 %	77 %
Deaths	47 %	0 %

DISCUSSION

Hemorrhagic infarction is present when there is extravasation of blood into an area of brain infarction such that blood and ischemic brain occupy the region of infarction. The presumed mechanism of formation of the HI is first occlusion of a brain artery producing ischemic damage to the brain and vascular endothelium followed by the re-establishment of blood flow. Blood then seeps through damaged vascular endothelium producing varying degrees of HI from small amounts of microscopic hemorrhage invisible to the

CT scan to larger quantities of blood visible as mottled areas of increased density on CT or as large confluent collections of hematoma within an infarct causing mass effect and clinical worsening. There seems little doubt that these latter large and often lethal brain hemorrhages are over represented in published studies by virtue of the fact that these patients are studied more extensively and are more apt to die thus becoming part of an autopsy series. Based on serial CT studies of patients with brain infarcts slightly fewer than 10 % of patients show evidence of HI (1). Of the HI's about 75 % will be associated with a cardiogenic brain embolus. The remaining 25 % occur with nonembolic brain infarcts. Others have found that only about one-third of patients with HI have a cardiac embolic source (2). Certainly evidence is mounting that the myth that the presence of an HI is always indicative of a cardiogenic cerebral embolus is incorrect. How could this belief have arisen? The answer rests in the fact that autopsy studies provided the first observations regarding HI. Seventy-eight' (3) percent to 95 % (4) of patients with HI's were felt to have suffered a cardiogenic cerebral embolus. Well over half of autopsied patients with a cardiogenic cerebral embolus were found to have a HI (3-5). Thus arose the concept that almost all HI's were the product of a cardiac derived embolus. Cardiogenic brain emboli are over represented in autopsy series because they generally produce larger more devistating strokes with a higher mortality (6-7). The present series would suggest that most HI's occur without worsening of symptoms and thus could easily go undetected as follow-up CT scans are often not routinely done. Forty-five (73 %) of the 62 cases of HI in this study occurred in the absence of clinical worsening. Based on the present study the most relevant factors in predicting a subgroup of cardioembolic stroke patients at risk for developing brain hemorrhage and worsening are the presence of a large cerebral infarct anticoagulated on the basis of a CT scan free of hemorrhage done soon, i.e., within 12 hours, after the embolic stroke. The management recommendations for a patient with a suspected cardioembolic stroke are to delay anticoagulation until a CT scan done between 24 and 48 hours from onset shows no hemorrhage. Delay anticoagulation for 7-10 days in patients with large embolic strokes and in patients with persistently elevated blood pressure, i.e., greater than 180/100. This practice avoids the risk of anticoagulation in a population of increased risk for spontaneous hemorrhagic transformation of their infarction.

REFERENCES

1. Fisher, M., Zito, J.L., Siva, A., DeGirolami, U.   Stroke 15: 192,1984.
2. Lodder, J.   Acta neur. scand. 70:329-335,1984.
3. Torvik, A., Jorgensen, L.   J. neurol. Sci. 3:410-432,1966.
4. Fisher, M., Adams, R.D.   J. Neuropath. exp Neurol. 10:92-93,1950.
5. Adams, R.D., Van der Ecker, H.M.   Annu. Rev. Med. 4:213-252,1954.
6. Yamaguchi, T., Minematsu, K., Choki, J.I., Ikeda, M.   Jap.
   Circul. J. 48:50-58,1984.
7. Sherman, D.G., Goldman, L., Whiting, R.B., Jurgensen, K.,
   Kaste, M., Easton, J.D.   Archs Neurol. 41:708-710,1984.

25

CEREBRAL EMBOLISM IN ATRIAL FIBRILLATION AND NONBACTERIAL THROMBOTIC ENDOCARDITIS

KURAMOTO, K., MATSUSHITA, S., YAMANOUCHI, H.

Tokyo Metropolitan Geriatric Hospital, Itabashi-KU, Tokyo 173, Japan

ABSTRACT

Pathogenetic role of atrial fibrillation (AF) and nonbacterial thromobotic endocarditis (NBTE) on cerebral embolism was studied in 2340 consecutive autopsies of the aged. Large brain infarcts in patients with AF were more often (21.9%) compared to infarcts in non-AF patients (7.3%). Intracranial cerebral artery athersclerosis was less severe in AF patients with large brain infarcts compared to non-AF patients with large infarcts. NBTE was found in 9.3% of consecutive autopsies and was associated with a high prevalence of large or medium sized brain infarct (49.7%) irrespective of co-existent disseminated intravascular coagulation, implicating embolism as the probable stroke mechanism.

INTRODUCTION

Atrial fibrillation is associated with an increased risk of brain infarction. In order to clarify the role of AF of the elderly to cerebral infarction, this clinicopathological study was carried out in 2340 consecutive autoposies of the aged. We also determined the prevalence of cerebral infarcts in NBTE and the relationship of NBTE with and without disseminated intravascular coagulation (DIC) to stroke.

METHODS

The subjects of the study were 2340 consecutive autoposies at Tokyo Metropolitan Geriatric Hospital from 1972 to 1982. They consisted of 1205 men and 1135 women over the age of 50. Atrial fibrillation was found in 405 cases (17.3%), in 17.8% of men and in 16.7% of women. The incidence of AF increased linearly with advancing age. The average age was 79.4 years in AF

(14.1%) and lung (13.0%).

The high prevalence (9.3%) of NBTE in this consecutive autopsy series is attributed to particularly careful examination of the heart, seeking these lesions at the autopsy table. The aortic valve was involved in 46.1%, the mitral valve in 40.6% and both aortic and mitral valves in 8.3%.

The cerebral infarctions of large, medium and small sizes were observed in 79.3% of NBTE and 61.4% of the control group (p 0.001) (Fig. 3). The incidence of large cerebral infarction was 14.7% in NBTE cases and 9.2% in the non-NBTE cases. The incidence of cerebral bleeding in NBTE patients was not different from patients without NBTE (Fig. 3).

DIC was found in 41.9% of NBTE, in 15.6% of the controls, and in 18% of the total cases. The incidence of cerebral infarction was only slighty higher in DIC (68%) than in non-DIC cases (62%). Thus, cerebral infarction seemed not to be strongly associated with intravascular coagulation from DIC, but supports embolism as the mechanism of stroke in NBTE.

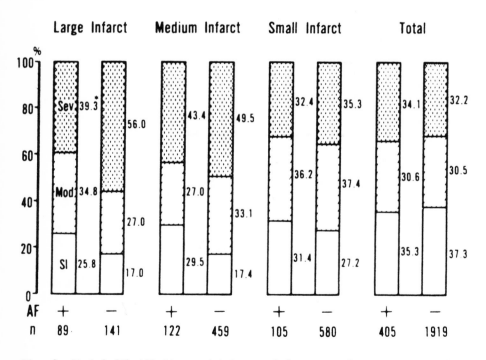

Fig. 2. Atrial fibrillation and intercranial atherosclerosis.

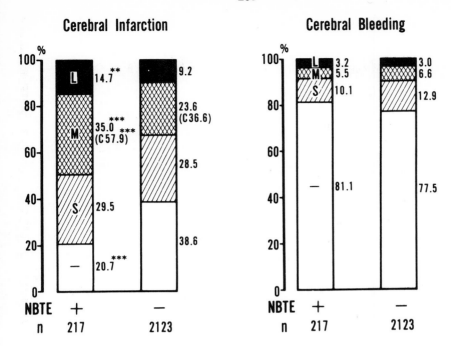

Fig. 3. NBTE and cerebral infarction and bleeding.
C: Cortical infarction in medium-sized infarction

REFERENCES

1. Kuramoto, K., Matsushita, S., Yamanouchi, H. Jpn. Circul. J.
   48: 67-74, 1984.
2. Kuramoto, K., Matsushita, S., Yamanouchi, H. Jpn. Circul. J.
   48: 1000-1006, 1984.

26

MITRAL AND AORTIC VALVE PROLAPSE IN YOUNGER PATIENTS WITH CEREBRAL ISCHEMIC EVENTS - RESULTS OF A PROSPECTIVE STUDY WITH TRANSTHORACAL AND TRANSESOPHAGEAL ECHOCARDIOGRAPHY

KRÄMER, G.(1), ERBEL, R.(2), TOPHOF, M.(1), MOHR-KAHALY, S.(2), ZENKER, G.(3)

(1)Department of Neurology and (2) Second Department of Internal Medicine, University Hospital of Mainz, FRG, (3) Second Department of Internal Medicine, LKH Graz, and Institute of Preventive Medicine Joanneum, Austria

ABSTRACT

Previous studies have revealed that mitral valve prolapse (MVP) is the most common risk factor for cerebral ischemic events in younger patients. In a prospective study using transthoracal two-dimensional echocardiography we found MVP in about half of more than 70 cases.

It is known that a prolapse of the aortic valve (AVP) may be associated with a MVP. Transthoracal echocardiographic diagnosis is however limited due to anatomical and technical reasons. On the other hand, an AVP seems to be even more relevant than a MVP with regard to possible cerebral embolic events from the pathophysiological point of view.

Therefore we decided to conduct additional transesophageal echocardiography in the patients of our study. Comparative results of the two methods are presented and discussed.

INTRODUCTION

Mitral valve prolapse is a very common condition with a prevalence of 6 to 12 percent in presumably healthy people. The majority of affected persons never develops any symptoms and needs no treatment. However, in 1974 Barnett and coworkers (1) recognized the association between cerebral and retinal ischemic events and mitral valve prolapse (MVP) in younger patients. In 1980, with one dimensional m-mode echocardiography, they were able to demonstrate a prolapsing mitral valve in 40 percent of 60 patients with cerebral ischemia younger than 45 years in comparison to only 6,8 percent of 60 age-matched controls (2). 18 of the 24 patients with MVP had no other detectable cause or risk factor.

In the meantime, other groups have published comparable results and in spite of an ongoing discussion due to the discrepancy of high MVP prevalence and low stroke incidence, the association of MVP and TIAs and strokes of young patients is now well established and accepted (3). The aim of our study was the determination of MVP prevalence in a consecutive patient-series with two-dimensional transthoracal and transesophageal echocardiography. In addition, we were interestd in the age-dependent role of MVP within the group of young patients and in patient subsets with an increased risk of cerebral embolization.

PATIENTS AND METHODS

68 patients with cerebral ischemic events below the age of 45 were included in a prospective study between December 1982 and 1984. All patients were examined by transthoracal two-dimensional echocardiography and 42 of them in addition had a transesophageal study at a follow-up examination. Platelet function was measured with levels of ß-thromboglobulin and platelet factor 4 in 57 patients. All had a detailed neurovascular diagnostic work-up including transfemoral cerebral angiography in 45 cases.

With transthoracal two-dimensional echocardiography a MVP was recognized in 32 of the 68 patients or 47 %. 19 of the patients with an abnormal valve were female and 13 male representing a ratio of 1.46 : 1. In contrast, only 15 of the 36 patients without MVP were female, a ratio of 0.71 : 1. The mean age in the two groups was 31,9 and 39 years respectively, with a range from 16 to 45 and 26 to 45 (for details see Table).

The known neurovascular risk factors showed a striking difference of distribution. 18 of the 32 patients with MVP had no known risk factor at all as opposed to only 2 of the 36 patients without MVP. When these 2 patients had their follow-up examination with transesophageal echocardiography a MVP could however be demonstrated.

Clinical features of the cases with an assumed relationship to MVP were weaker and more often in the form of a TIA or RIND in comparison to the patients without MVP. Disturbed platelet function could be demonstrated

Table 1

DIFFERENTIATION BETWEEN THE PATIENTS WITH AND WITHOUT MVP

	MVP (n = 32)	no MVP (n = 36)
FEMALE:MALE	19 : 13	15 : 21
AGE Mean	31,9	39,0
Range	16 - 45	26 - 45
RISK FACTORS		
hypertension	4	19
nicotine abuse	4	17
oral contraceptives	4 (of 19)	9 (of 15)
obesity	2	16
hyperlipemia	-	21
coronary heart disease	1	7
diabetes	-	3
miscellanous	2	8
NO RISK FACTOR	18	2
ISCHEMIC EVENTS		
TIA carotid	9	6
RIND carotid	3	4
vertebro-basilar	3	-
STROKE carotid	14	23
vertebro-basilar	3	3
PLATELET FUNCTION (mean values)		
ß-thromboglobulin	166,1	81,2
(normal up to 52 ug/l)		
platelet factor 4	74,9	30,3
(normal up to 10,4 ug/l)		
ANGIOGRAPHY (pathological findings)		
ICA stenosis	1	6
occlusion	-	4
MCA stenosis	1	-
occlusion	8	2
ACA occlusion	2	1
PCA occlusion	1	2

in both groups, but was much more pronounced in patients with MVP with a statistically significant difference (p-value of t-test below 0.01).

Cerebral angiography revealed in only one MVP-patient with a recent whiplash-injury an extracerebral ICA-stenosis; intracranial lesions of the middle cerebral artery as well as the anterior and posterior cerebral artery were more common. Again, in 4 of the 5 patients with intracranial vascular lesions and without MVP in transthoracal two-dimensional echocardiography a MVP was revealed by transesophageal follow-up examination. A comparison of two-dimensional transthoracal and transesophageal echocardiography in 42 of the 68 patients revealed 9 patients with negative transthoracal investigation. On the occasion of this follow-up examination, the transthoracal two-dimensional study was repeated by an independent second examiner. Comparison of the results showed a good reproducibility with only 3 cases of divergent results.

With the transesophageal examination, besides the improved detection of MVP, we were also able to demonstrate additional abnormalities like aortic and tricuspidal valve prolapse or aneurysms and septal defects in 18 patients. 20 of the 32 MVP patients had thickened mitral valves.

An analysis of the MVP-incidence in our patients showed a considerably varying frequency of MVP within the group of patients below 45 years. All our patients between 16 and 25 years and two thirds of those between 26 and 35 had a MVP. Thereafter, the prevalence declined to 45 % in patients between 36 and 40 years and 22 % in those between 41 and 45 years. These data clearly demonstrate an age-dependent declining probability of MVP-association with cerebral ischemia within the group of younger patients.

CONCLUSIONS

In conclusion, our results as well as those presented by others indicate that mitral valve prolapse is:
- The most common risk factor for cerebral ischemic events in younger patients.
- The incidence declines from very high values in the 16 - 25 year old group to about 25 % in the group between 41 and 45.
- Clinical diagnosis of MVP as well as sole m-mode echocardiography are relatively insensitive. At least a transthoracal two-dimensional

study with combined doppler echocardiography is necessary; transesophageal examination is a further improvement.
- Because of a possible coincidence, other causes of vascular disease must be ruled out.
- In most cases the clinical symptomatology of cerebral emboli is relatively mild and has a good prognosis.
- There seem to be patient subsets with MVP and increased risk of cerebral embolization, for instance those with:
    - thickening of the prolapsing mitral valve,
    - coincidence of aortic and tricuspidal valve prolapse, or
    - association with platelet dysfunction.

REFERENCES

1. Barnett, H.J.M., Jones, M.V., Boughner, D.R.   Trans. Am. Neurol. Ass. 100: 84-88, 1975.
2. Barnett, H.J.M., Boughner, D.R., Taylor, D.W., Coper, P.E., Kostuk, W.J., Nichol, P.M. New Engl. J. Med. 302: 139-144, 1980.
3. Boughner, D.R., Barnett, H.J.M. Stroke 16: 175-177, 1985.

27

INTERACTIVE VALUE OF CARDIAC ECHOCARDIOGRAPHY AND HOLTER RECORDING IN THE INVESTIGATION OF STROKE

BONNET, J., DESBORDES, P.(1), COSTE, P., CLEMENTY, J., ORGOGOZO, J.M.(1), BRICAUD, H.

Clinique Medicale Cardiologique, Hopital Cardiologique de Bordeaux, 33604 Pessac, France
(1) Unite de Pathologie Vasculaire Cerebrale, Hopital Pellergin, Place Amielie-Raba-Leon, 33076 Bordeaux cedex, France

Cardiogenic emboli have been implicated in a substantial percentage of episodes of cerebral and peripheral vascular ischemia and infarction. The identification of a cardiovascular cause is important in the management of stroke patients in order to outline an efficient plan of management.

Echocardiography and ambulatory electrocardiographic monitoring by Holter's technique are two noninvasive techniques capable of identifying a variety of potential cardiovascular causes of systemic embolism. A major goal of this study was to formulate guidelines for the use of echocardiography and Holter recording in patients with acute ischemic cerebrovascular disease. Another aim of our study was to outline the echocardiographic findings predictive of positive arrhythmia monitoring. Echocardiography can help to formulate the decision for Holter recording.

PATIENTS AND METHODS

This study is a retrospective one of 99 consecutive patients with thromboembolic stroke or transient ischemic attack (TIA). The population included 66 men and 33 women. The mean age was 55 + 12 years. No patient was over 75 years old. The majority of patients had completed ischemic cerebral infarction (n=79) and 20 had TIAs. The carotid territory was involved in 85 cases and the vertebrobasilar territory in 14 cases. The diagnosis of thromboembolic stroke was defined by CT scanner and by carotid or vertebrobasilar arteriography.

Cardiac assessment included (1) conventional M-mode echocardiography, (2) two dimensional echocardiography, and (3) arrhythmia monitoring by

Holter technique. However, these techniques are generally done only after a first evaluation including a physical examination, chest radiography and ECG.

A prior study (personal data) allowed us to define three ECG criteria which correlated with a positive arrhythmia monitoring study: (1) QRS axis under 45 % below 0, (2) presence of atrioventricular block and (3) presence of a Q wave. These criterias are used to distinguish a pathological or normal ECG.

RESULTS

Following the initial evaluation including the physical examination, chest radiography and ECG, we divided our patients into three groups. The first group (n=28) had no clinical evidence of cardiac disease. The second group (n=17) had systemic hypertension only, without cardiomegaly. The third group (n=54) had histories or findings sugggestive of cardiac disease. In this group the major cause of the initial event was atheromatous or hypertensive vascular disease with degenerative cardiomyopathy in 19 cases (56 %). Valvular disease was found in 12 cases.

Holter monitoring demonstrated an arrhythmia in 27 %. In most cases, a potentially prediposing arrhythmia was discovered: sinus bradycardia (n=9), atrioventricular dissociation (n=2), atrial ectopic beats of greater than 30 per hour (n=12), ventricular extrasystoles occurring more often than 30 per hour (n=14). In 11 cases a correlative arrhythmia was found including 7 cases of atrial fibrillation and 4 cases of supraventricular tachycardia.

If we consider these results in light of the classification after our initial cardiac investigation, we note that in the first group without cardiac disease and in the second group with isolated hypertension, the frequency of a positive Holter recording is very low, 10 and 12 % respectively in comparison to the third group where we found an arrhythmia in 48 % of cases. The analysis of individual cases shows in the first two groups only six cases of positive Holter recording with one case of atrial fibrillation and two cases of supraventricular tachycardia.

Thus, of 55 patients, 38 without clinically evident heart disease and 17 with lone hypertension, we found only 3 (5 %) significant arrhythmias, requiring antiarrhythmic treatment and/or anticoagulant drugs. Therefore, arrhythmia monitoring was of little value in patients with no clinical cardiac disease or with hypertension alone.

The echocardiograph is the second screening technique. Clinically adequate ultrasound imaging was obtained in 80 patients. In 64 % of cases, we found echocardiographic abnormalities, but no cases of atrial thrombi, mitral stenosis, cardiac tumor or infective endocarditis were identified. The more frequent abnormality was left atrial enlargement (telediastolic diameter larger than 38 mm, n=29, than 55 mm, n=17, 21 %). In 9 cases (11 %) mitral valve prolapse (MVP) was found and in 12 cases (15 %) ventricular hypertrophy. Of these 12 cases with ventricular hypertrophy only 7 patients had hypertension. We observed that in the group with clinical heart disease, echocardiographic abnormalities were found in 91 % of cases. In the group with isolated hypertension, 10 patients (61 %) had abnormalities that might predispose to thromboembolism, with left atrial enlargement (8 cases). On the contrary, in the group without heart disease, the abnormalities are rarer: 11 cases out of 20 patients (35 %). In this group without heart disease we found 3 cases out of 30 (10 %) of left ventricular enlargement.

The most important point in this group is the high frequency of MVP which was observed in 8 cases (22 %). This finding demonstrates the difficulty in diagnosing MVP by simple physical examination and ECG. However, these lesions should probably be considered only as abnormalities that might predispose to thromboembolisms and not as a direct cause.

When we correlated echocardiographic abnormalities with the arrhythmias discovered by Holter recording, we found a high frequency of left atrial and left ventricular enlargement, 55 % and 41 % respectively when Holter recording is positive, compared to a low frequency of these two abnormalities in the group of patients with negative Holter recording, 29 % and 14 % respectively. Left atrial size was significantly greater in the arrhythmic group compared to the group with a negative Holter recording (39.9 ± 8.11 mm versus 33.9 ± 6.9 mm, $p < 0.01$). Another point is the rarity of finding an arrhythmia by monitoring in MVP. In 9 cases of MVP, we found an arrhythmia only once, and in that case the arrhythmia was ventricular ectopic beats.

DISCUSSION

Our study shows that echocardiography in the population without clinically evident cardiac disease is of interest in detecting MVP but arrhythmia monitoring in these cases is most often negative. On the other hand, in lone hypertension the frequency of echocardiographic abnormalities is high consisting primarily of left atrial and ventricular enlargement. The discovery of these abnormalities may be important in that we showed them to be highly correlated with an arrhythmia.

Our results are in agreement with other studies (1-4) showing that there is little usefulness of echocardiography and arrhythmia monitoring in patients without obvious cardiac disease. In these patients without obvious cardiac disease echocardiography may be important for the discovery of MVP. However, the therapeutic decision for antiplatelet drugs is generally not modified by this discovery. Anticoagulants are likely to be reserved for cases of recurrent events. So the nonrecognition of MVP is probably not of great consequence after a first ischemic event.

Fig. 1   Management of cardiac exploration in patient with stroke.

STROKE

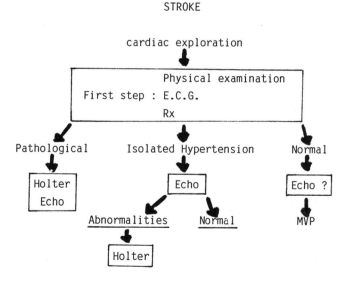

For lone hypertension an initial Holter monitoring study appears useless, but often echocardiography finds abnormalities predisposing to arrhythmias, and in these cases arrhythmia monitoring must be performed.

In conclusion, we think that the cardiac investigation in stroke must be organized in two steps (Figure 1). After the first step, if cardiac disease is clinically evident, Holter recording and echocardiography must be done. If lone hypertension is found, only echocardiography must be performed. If predisposing abnormalities are found, arrhythmia monitoring may be done. If lone hypertension is found, only echocardiography must be performed. If predisposing abnormalities are found, arrhythmia monitoring may be done.

In patients without cardiac disease, echocardiography and ambulatory electrocardiographic monitoring are of limited value and may be avoided. But in this case we must realize that we accept the nonrecognition of MVP in more than 20% of cases. If our therapeutic decision is not modified, we can accept this risk of nonrecognition, otherwise systematic echocardiography must be performed.

REFERENCES

1. Come, P.C., Riley, M.F., Bivas, N.K.   Ann. Neurol. 13: 527-531, 1985.
2. Knopman, D.S., Anderson, D.C., Asigner, R.W., Greenland, P., Mikell, F., Good, D.C.   Neurology 32: 1005-1011, 1982.
3. Babalis, D., Maisonblanche, P., Leclercq, J.F., Coumel, P.H.   Archs Mal. Coeur 77: 100-105, 1984.
4. Jackson, A.C., Boughner, D.R., Barnett, H.J.M.   Neurology 34: 784-787, 1984.

28

CARDIOEMBOLIC STROKE WORKSHOP

DISCUSSION OF ABSTRACTS (CHAIRMEN'S SYNOPSIS)

SHERMAN, D.G., HART, R.G.

Existing clinical data about cardioembolic stroke and its prevention are rudimentary and largely flawed. No meaningful consensus about many clinical issues can be reached at present. Currently available observational data should, however, serve to define important areas for prospective, well-designed studies. The discussion sessions of the workshop focussed on several recommendations pertinent to future studies of cardioembolic stroke.

1. There are no entirely reliable clinical criteria for the diagnosis of cardioembolic stroke. Future studies should carefully and specifically define their inclusion criteria and not vaguely refer to "clinical features compatible with...", etc. It was emphasized that many patients with potential cardioembolic sources have concomitant cerebrovascular disease. Careful definition of the criteria for diagnosis is mandatory, accepting the current diagnostic uncertainties.

2. Recurrent embolism to the brain, if involving the same arterial territory as the initial stroke, can be difficult to distinguish from worsening due to extension of an occult atherothrombotic infarct or due to hemorrhagic transformation. When reporting purported early recurrence of brain embolism, the site of second embolus should be stated. If it involves the same arterial territory as the initial embolism, the reason for suspecting recurrent embolism versus other mechanisms of worsening should be clearly stated.

3. Embolism to the systemic (nonbrain) circulation are similarly difficult to clinically diagnose, and should be specifically defined. In some clinical reports, macroscopic hematuria has been equated with renal embolism. This approach is unsatisfactory.
   For example, renal embolism follows the brain as the most common site

in autopsy studies, reflecting the distribution of cardiac output. The presenting features of renal embolism are nonspecific, and include abdominal or flank pain, nausea, and low-grade fever. Small, segmental renal infarcts are often completely asymptomatic. Microscopic hematuria and mild proteinuria are found in only one-third of such patients. Elevation of serum creatinine does not occur unless both kidneys are involved. Marked elevation of serum lactate dehydrogenase is usually seen, and although nonspecific, is a helpful clue that renal infarction has occurred. While an intravenous pyelogram or isotope renogram may establish the diagnosis of large renal infarcts, renal artery arteriography is required to diagnose small emboli. The clinical diagnosis of emboli to the lower extremities can be difficult. If crural arterial occlusion occurs in older patients, coexistent atherosclerotic peripheral vascular disease may be difficult to discount, although less likely to produce ischemia of abrupt onset without prior symptoms.

4. Blindedness is critical in scientific validation of new diagnostic techniques. For example, the presence or absence of mitral valve prolapse by echocardiography is heavily influenced by individual interpretation (Wann et al. American Heart Journal 109: 803-808, 1985) and the likelihood of investigator bias is high in nonblinded studies of mitral valve prolapse and ischemic stroke. Blindedness is essential.

Several investigators challenged the generally accepted concept that hemorrhagic infarct is more common in cardioembolic stroke than in stroke from other mechanisms. An alternative hypothesis suggests that hemorrhagic infarcts are associated with large infarcts, and that large infarcts may be more common in cardioembolic stroke. Hence, infarct size and not stroke mechanism may be the determinant of hemorrhagic tendency.

Depending upon the cardiac source, emboli can be composed predominantly of platelets (nonbacterial thrombotic endocarditis), laminated red, fibrin-rich thrombi (ventricular aneurysm), calcific material (aortic stenosis), or even tumor (myxoma). Antithrombotic therapy that is effective for one cardioembolic source cannot be assumed to be effective in another.

INDEX